T0257630

Encyclopedia of Alternative and Renewable Energy: Trends and Applications of Renewable Energy

Volume 01

Encyclopedia of Alternative and Renewable Energy: Trends and Applications of Renewable Energy Volume 01

Edited by **Craig Zodikoff and David McCartney**

New York

Published by Callisto Reference,
106 Park Avenue, Suite 200,
New York, NY 10016, USA
www.callistoreference.com

Encyclopedia of Alternative and Renewable Energy:
Trends and Applications of Renewable Energy
Volume 01
Edited by Craig Zodikoff and David McCartney

International Standard Book Number: 978-1-63239-175-9 (Hardback)

Printed in the United States of America.

Contents

Preface

This book elaborately discusses the various trends as well as applications of renewable energy. Rise in electricity demand and environmental issues have led to the rapid development of energy production from renewable resources. In the long run, application of renewable energy sources (RES) can assure ecologically sustainable energy supply. This book presents recent advancements and developments in renewable energy resources distributed over numerous topics. It can be a source of information and basis for discussion for readers from diverse backgrounds. The book includes contributions of renowned experts, scientists and researchers in this field and is aimed towards providing significant information about renewable energy sources and their applications.

After months of intensive research and writing, this book is the end result of all who devoted their time and efforts in the initiation and progress of this book. It will surely be a source of reference in enhancing the required knowledge of the new developments in the area. During the course of developing this book, certain measures such as accuracy, authenticity and research focused analytical studies were given preference in order to produce a comprehensive book in the area of study.

This book would not have been possible without the efforts of the authors and the publisher. I extend my sincere thanks to them. Secondly, I express my gratitude to my family and well-wishers. And most importantly, I thank my students for constantly expressing their willingness and curiosity in enhancing their knowledge in the field, which encourages me to take up further research projects for the advancement of the area.

Editor

Are Renewables Effective in Promoting Growth? Evidence from 21 EU Members

António C. Marques and José A. Fuinhas

NECE, University of Beira Interior, Management and Economics Department, Covilhã, Portugal

1. Introduction

The energy needs of mankind in the 19th century were essentially satisfied by the use of renewable sources, such as biomass mainly by burning wood, and animal power. Some transformation of agricultural goods was done through the exploitation of natural resources such as wind and water, using wind mills and water wheels. The 20th century was the century of high economic growth. It was a century where the use of internal combustion engines was widespread and with them the massive use of fossil fuels. The 21st century is now looking for alternative sources of energy.

Nowadays, there is a backdrop of the forecast of depleting fossil fuels in the near future, particularly oil, and climate changes, associated with large emissions of carbon dioxide. In this century, there is a great focus on renewable energy sources, with the strong support of public policies. In addition to the use of hydropower, which already has mature technology, there is a continuous process of developing technologies for harnessing the wind and photovoltaic resources. The attribution of the title of energy of the future to renewable sources is dependent on two factors. First, the achievement of their own economic sustainability will depend on the evolution of technology itself. Second, it will depend on the long-term advances in atomic energy to accomplish the nuclear fusion process on Earth.

One of the fundamental questions that arises is to assess whether this progressive change in the energy paradigm will affect the process of economic growth. The analysis of the relationship between energy consumption and economic growth is far from new in literature. Narayan & Smyth (2008) summarise the principal achievements, the absence of consensus and the diversity of methodologies. The study of the impact of using renewables on economic growth is, however, scarce (e.g. Apergis & Payne, 2010 & Menegaki, 2011). Furthermore, the literature has not focused on the energy mix, that is, on the impact of the simultaneous use of different energy sources on economic growth. Will the impact of energy on economic growth be identical, regardless of whether this energy comes from fossil fuels or renewable sources?

The literature is not unanimous regarding the relationship between income and environmental concerns. Some authors, such as Vachon & Menz, 2006 and Huang *et al.*, 2007, argued a positive effect of wealth on renewables. On the one hand, higher income

[1]Research supported by: NECE, I&D unity funded by the FCT - Portuguese Foundation for the Development of Science and Technology, Ministry of Science, Technology and Higher Education.

could cushion the effect of the greater fees and costs resulting from the encouragement of Renewable Energy (hereafter RE). On the other hand, higher income levels can also represent a further financial capacity for accommodating the huge investments needed to develop the technology of energy production from renewable sources. Recently, Marques *et al.* (2011) pointed out that the effect of Gross Domestic Product (GDP) on renewables is far from uniform. In fact, this effect depends on the phase of deployment of renewables. With the exception of countries that use renewable sources on a small scale, the authors show that in general the effect of GDP on RE is negative.

Higher income levels of countries have historically been associated with the use of technology based on fossil fuels. Currently, the production structures are grounded firmly in the use of internal combustion engines, in gas turbines and more recently in fuel cells, although the latter with still little relative weight. The technology associated with these fuel cells is still evolving. Right now, this technology is very expensive and partly controversial, because hydrogen is not a primary energy source. The process for obtaining it can lead to the release of harmful gases into the atmosphere. In addition, unequivocal confirmation is still lacking as to their efficiency when compared to that of conventional combustion engines. The conversion and replacement of all these technologies with other technologies based mainly on electricity, such as electric motors, is a long and extremely costly path. More often than not, this shift to alternative technologies is not desired or it is hindered by entrenched interests, which are mostly associated with lobbying exercised for players acting in traditional sources of energy.

There are specific features identified in renewable energies. The first refers to the mechanism of price formation. Given that the contracts of exploitation of renewable sources are generally long-term, usually with prices that are defined or indexed to inflation, the volatility of those energy prices that economic agents face is well controlled, when compared with the prices of fossil sources. This mechanism may thus contribute to increased stability of the economic environment in which agents conduct their forecasts and make their investment decisions. In this way, economic growth could be stimulated. Economic growth can also be promoted by the development of a cluster associated with renewables, since production is done locally. The use of local resources is just another feature of renewables.

The third feature concerns the incentives for the development of renewables. Countries have adopted an extensive battery of measures, such as feed-in-tariffs, grants or preferential loans, to encourage the development of these sources. Walking the path of renewable technology development requires sustaining high investment costs. In one way or another, the costs of implementing these policies are passed on to the economic agents. Since resources are scarce, if the inputs become more expensive, the focus on renewables may create inefficiencies in the economy. Inefficiency can also result from the fact that greater use of RE can cause already installed production capacity to be left behind, including capacity associated with internal combustion engines. Actually, these characteristics sum up the hot debate about the benefits / need for development of renewables. Therefore, to shed some light on this debate, we consider it indispensable to test empirically the effect that different energy sources have had on economic growth.

Since the beginning of the discussion about climate change, especially since the United Nations Framework Convention on Climate Change in 1990, and the Kyoto Protocol (1997), Europe has been firmly committed to this goal. That is why, within a context of energy policy and in order to promote the use of renewables, Europe has produced strong compulsory recommendations for its Member States. In the context of the definition of climate and energy targets, to be reached in 2020, the European Union (EU) established the 20-20-20 strategy. This strategy pursues the following objectives: i) reduce greenhouse gas emissions to at least 20% below 1990 levels; ii) produce 20% of energy consumption from renewable sources; and iii) encourage energy efficiency, reducing primary energy use by 20%. Plans are still being drawn up to make the target reduction of greenhouse gas emissions even more ambitious. Regarding the analysis of the relationship between sustainable development and economic growth, it is therefore important to study the EU region. In parallel with the clear commitment to extend the use of renewables, Europe has undergone growth difficulties. The European Council in Lisbon in the year 2000, and the Spring European Council in Brussels in 2005 defined, as their main goals, sustainable economic growth and job creation.

2. The debate on economic growth within a context of energy paradigm change

All economic growth has a unique framework and, as such, it must be considered as a result of a whole. As far as the relationship between primary energy sources and economic growth is concerned, the literature assesses four main hypotheses. First, when there is a unidirectional causality running from energy to economic growth, we are in the presence of the growth hypothesis. This implies that economic growth requires energy, and as a consequence, a fall in primary energy consumption is likely to hamper economic growth. Second, once a unidirectional causality running from economic growth to energy is established, then the conservation hypothesis is verified. This means that economic growth is not totally dependent on energy consumption and therefore few or negligible effects on economic growth are expected from energy conservation policies. Third, the bidirectional causality between energy and economic growth is known as the feedback hypothesis. In other words, the rise in primary energy demand provokes economic growth and vice versa. Finally, the neutrality hypothesis sustains that policies on energy consumption have no consequences on economic growth, due to the neutral effect with respect to each other.

Indeed, economic growth could be influenced by several factors that ultimately determine its performance. The energy it uses as input, the energy dependence in relation to the outside and the volatility of its own process of evolution are driving forces in this economic growth path. Energy is traditionally identified as a key driver of economic growth but, in fact, it is unlikely that all sources of energy produce the same impact. Their different characteristics, such as the cost /benefits balance, environmental consequences, state of maturation of the technology and even their scale of production can determine the effect of each of these sources in the dynamics of the economic growth process.

2.1 Energy sources, external dependency and economic growth

The literature focusing on the relationship between energy consumption and economic growth is vast and diverse. Some studies focus on the reality of particular countries (Lee & Chang, 2007; and Wolde-Rufael, 2009), while others centre on groups of countries (Akinlo, 2008; and Chiou-Wei et al., 2008). Most of them are engaged in the study of the direction of causality, both in the short and long run. The recent papers of Odhiambo (2010), Ozturk (2010), and Payne (2010) are good surveys. The empirical literature on causality between RE and economic growth has achieved mixed results. For twenty OECD countries, Apergis & Payne (2010) estimated a panel vector error correction model, and found bidirectional causality between RE and economic growth, both in the short and long run. A bidirectional causality between RE and economic growth was also detected by Apergis *et al.* (2010), for a group of 19 developed and developing countries. For the US, Menyah & Wolde-Rufael (2010) found only a unidirectional causality running from GDP to RE. Conversely, when it comes to analysing the relationship between RE and economic growth, the empirical literature is thin. Menegaki (2011) is one of the exceptions, studying the situation in Europe. Indeed, focused on 27 EU Members, using panel error correction, the author did not confirm the presence of Granger causality running from RE to economic growth, either in the short or long run. These results lead the author to conclude that the consumption of RE makes a minor contribution to GDP. In fact, it seems that the nature of the relationship between RE and economic growth still has a long way to go before consensus is achieved.

The literature on the empirical link between restraining emissions of carbon dioxide (CO_2) and economic growth has shown some unexpected results. Menyah & Wolde-Rufael (2010) only found unidirectional causality running from CO_2 to RE. In the same way, Apergis *et al.* (2010) conclude that the consumption of RE does not contribute to reducing CO_2 emissions. Their explanation is grounded in the well-known problem of storing energy, as well as the intermittency characteristic of renewables. The failure to store energy, for example from wind or solar sources, requires the simultaneous use of established sources of energy, such as natural gas or even the highly polluting coal. This scenario leads to two effects on the installed capacity and on energy dependency. On the one hand, it implies the maintenance and even the enlargement of productive capacity that becomes idle for long periods, which generates economic inefficiencies. On the other hand, the intermittency may not even contribute to the reduction of a country's energy dependence goal, such as documented by Frondel *et al.* (2010).

The root of the lack of consensus in literature, with regard to the relationship between RE and economic growth, could come from different theoretical and practical perspectives. On the one hand, it is admissible that the effect of RE on economic growth could vary largely according to both the geographical area and the time span analysed. On the other hand, there could be a variable omission bias problem. In fact, the research may be disregarding the importance of other variables, such as the simultaneous consumption of oil, coal, nuclear or natural gas. These variable omissions could lead to wrong conclusions on causality between each energy source and economic growth, when analysed separately. The cost of this is that inconsistent and erroneous results may be achieved.

Under the well-known premise that energy plays a crucial role in the economic growth process, the question that arises is what will the particular role of renewable sources of energy be on economic growth? To find the answer to this question, as stated before, possible bias resulting from the omission of variables must be avoided, and it is necessary to assess the simultaneous explanatory power of the main sources of energy driving economic growth.

2.2 Volatility and economic growth

The problem of GDP growth analysis has a long path in economic literature. The mainstream does not sustain any relationship between economic growth and its volatility. Nevertheless, the relationship between economic growth volatility and the trend in growth has been the object of increasing attention in literature. Indeed, macroeconomists have long focused on business cycles and economic growth. The material progress of humankind is a central issue. As an economic problem, however, it has been a very complex matter. Although the core literature does not advance any reason for volatility exerting a specific effect on economic growth, this statement is not true for all authors. Some authors advance that volatility might have a reducing effect on economic growth, while others suggest that a positive effect may be observed on economic growth. In short, the literature indicates that the relationship between volatility and economic growth may be: 1) independent – the mainstream; 2) negative - e.g. Bernanke (1983), Pindyck (1991), Ramey & Ramey (1995), Miller (1996), Martin & Rogers (1997 and 2000), and Kneller & Young (2001); or 3) positive - e.g. Mirman (1971), Kormendi & Meguire (1985), Black (1987), Grier & Tullock (1989), Bean (1990), Saint-Paul (1993), Blackburn, (1999), and Fountas & Karanasos (2006). A good survey on this relationship was undertaken by Fang & Miller (2008).

Several explanations have been advanced to support this controversial link. The negative relationship could come from several paths. Volatility limits investment, which limits demand and therefore constrains economic growth (Bernanke, 1983; and Pindyck, 1991). At the same time, volatility can be harmful to human capital accumulation, which diminishes economic growth (Martin & Rogers, 1997).

A positive association between economic growth and volatility could result from diversified sources. The volatility of economic growth generates high precautionary savings (Mirman, 1971). Further specialisation tends to coexist with further economic growth volatility, as highly specialised technologies only generate investment if their expected returns are high enough to compensate for the risk (Black, 1987). This latter assumption suggests that bursts in volatility tend to be related to high economic growth. The cost of opportunity related to productivity-improving processes tends to drop in recessions resulting in a positive relationship between economic growth volatility and economic growth (Bean, 1990; and Saint-Paul, 1993). Labour market institutions, the technology of production, and the source of shocks are characteristics that increase the pace of knowledge accumulation, lowering economic growth (Blackburn, 1999; and Blackburn & Pelloni, 2004). Higher economic growth leads to higher inflation, in the short run, according to the Phillips curve approach (Fountas & Karanasos, 2006).

The explanations for economic growth volatility point out that on the one hand, monetary shocks generate economic growth fluctuations around its natural rate that reflect price misperceptions. On the other hand, technology and other real factors influence the long-run economic growth rate of potential economic growth. These two approaches consubstantiate the misperception theory of Friedman (1968), Phelps (1968), and Lucas (1972). The expectations about economic growth volatility could exert an impact on economic growth (Rafferty, 2005). In sum, the mix of results strongly suggests that theoretical and empirical developments are required to establish the nexus between economic growth and volatility growth.

3. European Union

For a long time now, the EU has taken the lead in the fight against climate change. As stated before, one of the tools that the EU has used is to set targets for the use of RE, in each of the EU Member States. Some of the important milestones along the way have been the White Paper for a Community Strategy and Action Plan, Energy for the Future: Renewable Sources of Energy", in December, 1997 and the EU Directives 2001/77/EC and 2009/28/EC. In parallel with concerns about climate change, concerns are also emerging in Europe about economic growth, which has been generally modest.

3.1 Current picture

The commitment to renewables made by the EU has been translated into real achievements with regard to the contribution of these sources to the energy supply. For the period 1990-2007, and for the EU of 27, we looked at the picture of the evolution of GDP growth rates, as well as the evolution of the contribution of renewable sources to total energy supply (CRES), as a percentage.

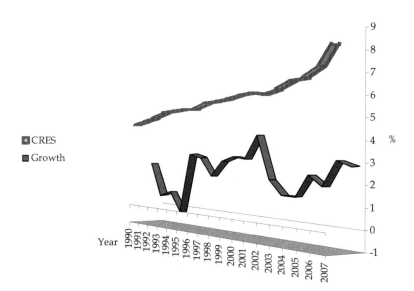

Fig. 1. Economic growth and renewable energy use in EU27

Figure 1 suggests that the rising trend in the use of RE is contemporaneous with different behaviours of economic growth. The periods of greatest growth in the use of renewable sources were simultaneous with contractions in economic growth. This gap is clearly visible in the mid-1990s and the 2000s, and the CRES variable clearly accelerates during 2000s.

When we analyse this reality in detail, we find that data is missing for some of the 27 countries, in particular with regard to variables related to the use of other energy sources. Thus, the EU Members for which the data is available, for all the variables considered and for the time span under review, are: Austria, Belgium, the Czech Republic, Denmark, Estonia, Finland, France, Germany, Greece, Hungary, Ireland, Italy, Luxembourg, the Netherlands, Poland, Portugal, the Slovak Republic, Slovenia, Spain, Sweden, and the United Kingdom. For the time span 1990-2007, this chapter is focused on this panel of 21 EU Members. For this panel, we made a first inspection by country, into the relationship between the growth in renewables' use and economic growth. To do so, we calculated both the average rate of the growth rates of contribution of renewables to total energy supply, and the average rate of economic growth in the period 1991-2007. This information is summarised in Figure 2.

In general, it was observed that the highest rates of growth in the use of RE are associated with countries with lower economic growth. The highest average rate of economic growth during this period was found for Ireland. For this country, the rate of growth in the use of renewables (4.22%) was markedly lower than that rate of economic growth (6.48%). The highest rates of growth in the use of renewable sources in this period are usually associated with countries that have shown lower economic growth rates. Estonia, the Slovak Republic and the Czech Republic have the highest average growth rates of the use of renewables (nearly 12.1%, 11.3% and 9.5%, respectively), but they have low rates of economic growth (2.82%, 2.88% and 1.99%, respectively). Note that the average economic growth rate is 2.76% and the average growth of use of renewables is 5.17%. Germany has one of the highest growth rates of renewables (8.7%) during this period, but its average economic growth was only 1.58%.

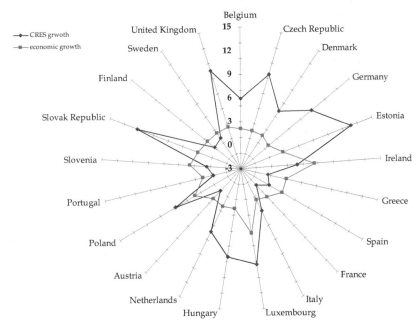

Fig. 2. Economic growth and renewable energy use

3.2 Variables
This chapter is focused on analysing the relationship between economic growth and the use of different sources of energy. We define as dependent variable the Logarithm of real Gross Domestic Product for country c, at period t, ($LGDP_{ct}$). The explanatory variables arise from the literature and are in accordance with those previously discussed. Therefore, in addition to the energy sources, we also control for energy consumption per capita, dependency on foreign energy and economic growth volatility. We then present and discuss the variables, their measurement and the expected contributions to economic growth. Given that volatility is a built variable instead of an observable one, we will explain in detail the process followed for its computation.

- *Per capita energy consumption (ENERGPCct)*. Energy consumption could be used either as a development or an energy efficiency indicator (e.g. Toklu *et al.*, 2010). We can observe two different effects with regard to the relationship between energy consumption and economic growth. On the one hand, higher energy consumption can lead to an increase in the use of the installed capacity. Therefore, larger consumption could stimulate production, and thereby boost economic growth. This yields a positive effect of energy consumption on economic growth. On the other hand, a negative effect can be observed, which can result from two phenomena: i) the energy is consumed in activities other than production; and ii) the increase in consumption also increases the cost of energy and is likely burdening the foreign energy deficit. The expected sign is negative if the former effect prevails.
- *Import dependency of energy (IMPTDPct)*. We control for the external energy dependence, which is often pointed out by normative literature as one of the major constraints on economic growth. Since the entire economy is closely linked to energy, the external dependency of that input not only causes huge capital flows to the outside, but also positions the country as a price-taker in the international energy markets. Thereby, we tested the hypothesis that the dependency on energy imports is limiting economic growth.
- *Per capita GDP volatility (VOLGDPPCct)*. Volatility is not a directly observable variable. It has the additional problem of the coexistence of several definitions. To cope with these complexities, we use autoregressive conditional heteroskedasticity (GARCH) models. The GARCH models are profusely used due to their recognised ability to capture many properties of time series, such as time-varying volatility, persistence and volatility clustering. In particular, GARCH processes have often been used in empirical literature to compute risk.

GDP growth has often been modelled as an autoregressive time series with random disturbances having conditional heteroskedastic variances. GDP growth, in particular, has been modelled as a GARCH type processes. The GARCH model is, in effect, sufficient to allow different macroeconomic regimes by letting the volatility of the economic growth evolve over time. It also assumes that a large change in GDP growth, either positive or negative, is probably followed by other large changes in subsequent years. Other methods of computing volatility, such as variance (or standard deviations), imply loss of observations and have several handicaps. Alternatively, they treat positive or negative changes in some way (the squares of economic growth rates) and were therefore excluded from our analysis.

We fit an autoregressive (AR) process with GARCH errors to the natural logarithm of the GDP per capita growth rates, assuming that the distribution of the error process is the normal distribution. This option results from the well-known characteristics of persistence of GDP growth. Indeed, we begin with the simplest model, namely the AR(1)-GARCH (1,1). Given the well-known propensity of the GARCH model to generate high estimated values at the beginning and end of time span, the AR(1)-GARCH(1,1) was estimated using raw data for the time span of 1971 to 2010.

The model estimated has a mean equation and an equation for conditional variance, which are, respectively (1) and (2):

$$DLGDPPC_t = \gamma_0 + \gamma_1 DLGDPPC_{t-1} + \varepsilon_t , \qquad (1)$$

and

$$\sigma_t^2 = \omega + \alpha \varepsilon_{t-1}^2 + \beta \sigma_{t-1}^2,\tag{2}$$

where ε_t is the error term. In the above model, equation (1) is the conditional mean equation and equation (2) is the conditional variance equation. The conditional standard deviation term, σ_t, represents the measure of GDP per capita growth volatility. One can also view σ_t as a measure of economy wide risk.

Since we are more interested in the level of volatility than in the volatility itself (σ_t), we proceed to establish the trend of volatility (*VOLGDPPCct*) applying the well-known Hodrick & Prescott (1997) – HP filter to the volatility obtained from the AR(1)-GARCH(1,1). Following a standard procedure of the related literature on HP filter, we use the value of λ =100 as the smoothing parameter.

Figure 3 shows the computed trend volatility. In general, there is no uniform behaviour pattern for the countries. For the time span analysed we observe the three possible kinds of trend: increase, decrease, and stability. For example, Austria and Spain reveal a period of stability until the end of the 1990s and a marked decline thereafter. In their turn, countries like Ireland, Luxembourg, and Poland show a trajectory of declining volatility. On the contrary, countries like France and Hungary reveal an increasing path with regard to volatility.

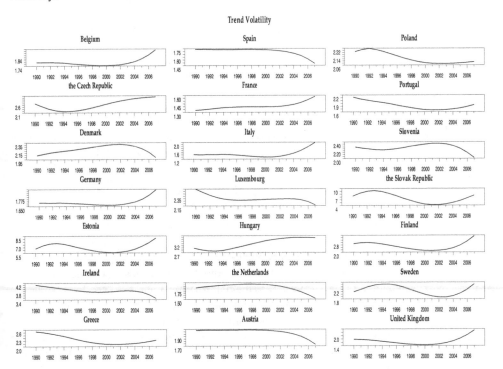

Fig. 3. Volatility trend

- *Logarithm of the contribution of renewables to total primary energy supply, lagged one period (LCRESct-1).* As discussed earlier, it is well known that economic growth is heavily dependent on energy use. Therefore, the contribution of each source towards economic growth should be assessed. Although renewables have yet to play a leading role in the total picture of energy sources in most countries, the relationship between renewables and economic growth must be evaluated. In reality, we are witnessing a growth rate of this source, largely as a result of public policies. On the one hand, these market opening policies or market driven policies take time to produce the desired effects and, on the other hand, the present productive structures are mostly suitable for the use of traditional sources. Thus, we control for the logarithm of the contribution of renewables to total primary energy supply, lagged one period. The effect of *LCRESct-1* can evolve in two directions. On the one hand, greater use of renewables may encourage the development of this entire industry, creating jobs and wealth locally. In this scenario, we will have a positive effect. On the other hand, greater use of renewables may involve the abandonment of fossil-based productive capacity and, therefore, we can observe a negative effect of renewables on economic growth. If the cost of the market-opening policies is excessively placed on the economy, then this negative effect can also be enlarged. If the second effect overcomes, then a negative signal is achieved.
- *Contribution of coal, oil, gas, and nuclear to electricity generation (SCOALEGct, SOILEGct, SGASEGct, and SNUCLEGct).* The conventional energy sources, including both fossil fuels and nuclear energy, are the dominant sources of energy and, as such, we control for the effect of all these sources on economic growth. Since the production structures in Europe are geared mainly towards the use of oil, we anticipate a clear positive effect for this source on economic growth. The same is expected to happen with nuclear power. With regard to coal and natural gas, given that the former source is highly inefficient and the latter is relatively recent, the expected effect may not be obvious *a priori.*

3.3 Method

This chapter makes use of panel data techniques to assess the nature of the effects of the several energy sources, and other drivers, on economic growth. Complex compositions of errors could be present in panel data analysis. The general model to estimate is:

$$LGDP_{ct} = \alpha + \delta LCRES_{ct-1} + \sum_{k=1}^{k} \beta_k X_{kct} + d_c + d_t + \mu_{ct}, \tag{3}$$

where $LCRES_{ct-1}$ is the share of renewables of country c in period t−1. The dummy variables d_c and d_t refer to country and time, respectively. In the error term $\mu_{ct} = \rho_c \mu_{c,t-1} + \eta_{ct}$, η_{ct} is serially uncorrelated, but correlated over countries.

To deal with the complexity of the errors, good econometric practices suggest performing the analysis by first making a visual inspection of the nature of the data, followed by a battery of tests to detect the possible presence of heteroskedasticity, panel autocorrelation, and contemporaneous correlation. We use the Modified Wald test (Baum, 2001) in the residuals of a fixed effect regression, to appraise the existence of groupwise heteroskedasticity. The Modified Wald test has χ^2 distribution and tests the null of: $\sigma_c^2 = \sigma^2$, for $c = 1,...,N$. The Wooldridge test assesses the presence of serial correlation. It is normally distributed $N(0,1)$ and it tests the null of no serial correlation. We use the parametric testing procedure proposed by

Pesaran (2004), the non-parametric test from Friedman (1937) and the semi-parametric test proposed by Frees (1995 and 2004), either for fixed effects or random effects, to test the countries' independence. Pesaran's test is a parametric testing procedure and follows a standard normal distribution; Frees' test uses Frees' Q-distribution; Friedman's test is a non-parametric test based on Spearman's rank correlation coefficient. All these tests - Pesaran, Frees and Friedman - test the null of cross-section independence.

Within a panel data analysis, the presence of such phenomena discourages the use of the common Fixed Effects (FE) and Random Effects (RE) estimators, due to the inefficiency in coefficient estimation and to biasedness in the estimation of standard errors they could cause. In this case, the appropriate estimators to be used are the Feasible Generalised Least Squares (FGLS) and the Panel Corrected Standard Errors (PCSE). In our sample, the number of cross sections (21) is larger than the number of time periods (18) and, therefore, the best suited estimator to deal with the presence of panel-level heteroskedasticity and contemporaneous correlation is the PCSE (Reed & YE, 2009).

The PCSE estimator allows the use of first-order autoregressive models for μ_{ct} over time in (3), it allows μ_{ct} to be correlated over the countries, and allows μ_{ct} to be heteroskedastic (Cameron and Triverdi, 2009). We begin by estimating a pooled OLS model (model I) and then we work on a panel data structure by applying the PCSE estimator. We will estimate the model presupposing the various assumptions about variances across panels and serial correlations, with the aim of checking the robustness of the results. The assumptions made throughout the models are as follows: model II - correlation over countries and no autocorrelation; model III – country-level heteroskedastic errors and common first-order autoregressive error (AR1); model IV - correlation over countries and autocorrelation AR(1); and model V - correlation over countries and autocorrelation country-specific AR(1).

3.4 Data
The data used in this chapter come from several sources. Table 1 summarises the variables, their sources and their descriptive statistics. The time span is 1990-2007, and we collect data for 21 EU Members, those for which there are available data for all the variables.

Variable	Definition	Source	Obs	Mean	SD	Min	Max
Dependent							
$LGDP_{ct}$	Logarithm of real Gross Domestic Product (billion dollars, 2005)	World Bank World Development Indicators, and International Financial IMF Statistics	378	5.3867	1.4966	1.9095	7.9921
Independent							
$ENERGPC_{ct}$	Per capita energy (kgoe/cap)	EU Energy in Figures 2010 DG TREN	378	4062.822	1590.981	1753.7	10132.98

Variable	Definition	Source	Obs	Mean	SD	Min	Max
$VOLGDPPC_{ct}$	*Per capita* GDP volatility	Own calculation. Raw data from World Bank World Development Indicators, and International Financial Statistics of the IMF	378	2.5407	1.2422	1.0622	8.7522
$LCRES_{ct-1}$	Logarithm of the factor of contribution of renewables to total primary energy supply, lagged one period	OECD Factbook 2010	376	1.5965	1.0126	-1.6094	3.4404
$IMPTDP_{ct}$	Import dependency of energy (%)	EU Energy in Figures 2010 DG TREN	378	52.2925	29.6911	-50.83	99.8
$SCOALEG_{ct}$	Contribution of coal to electricity generation	Ratio electricity generation to coal (TWh) / total elect. generation (TWh). EU Energy in Figures 2010 DG TREN	378	0.3614	0.2753	0	0.97
$SOILEG_{ct}$	Contribution of oil to electricity generation	Ratio electricity generation to oil / total elect. Generation. EU Energy in Figures 2010 DG TREN	378	0.0698	0.0983	0	0.51

Variable	Definition	Source	Obs	Mean	SD	Min	Max
$SGASEG_{ct}$	Contribution of gas to electricity generation	Ratio electricity generation to gas / total elect. Generation. EU Energy in Figures 2010 DG TREN	378	0.1694	0.1747	0	0.76
$SNUCLEG_{ct}$	Contribution of nuclear to electricity generation	Ratio electricity generation to nuclear / total elect. Generation. EU Energy in Figures 2010 DG TREN	378	0.2126	0.2306	0	0.78

Table 1. Data: definition, sources and descriptive statistics

First following a visual inspection of the data, we analyse the correlation coefficients, which are disclosed in the correlation matrix (table 2). In general, the correlation coefficients did not arouse any particular concern about the existence of collinearity among explanatory variables, although the correlation of $VOLGDPPC$ with $LGDP$ may be a possible exception.

Variables	$LGDP_{ct}$	$ENERGPC_{ct}$	$VOLGDPPC_{ct}$	$LCRES_{ct-1}$	$IMPTDP_{ct}$	$SCOALEG_{ct}$
$LGDP_{ct}$	1					
$ENERGPC_{ct}$	-0.1478	1				
$VOLGDPPC_{ct}$	-0.6610	-0.0209	1			
$LCRES_{ct-1}$	-0.0332	-0.0919	-0.1471	1		
$IMPTDP_{ct}$	-0.1230	0.1585	0.0574	0.0838	1	
$SCOALEG_{ct}$	-0.2211	-0.4187	0.1621	-0.1871	-0.4832	1
$SOILEG_{ct}$	0.1553	-0.4307	-0.1612	0.0342	0.3339	-0.0579
$SGASEG_{ct}$	0.1260	0.3487	-0.1024	-0.3672	0.1555	-0.3434
$SNUCLEG_{ct}$	0.1895	0.1240	0.0889	0.0640	0.0151	-0.4177

	$SOILEG_{ct}$	$SGASEG_{ct}$	$SNUCLEG_{ct}$
$SOILEG_{ct}$	1		
$SGASEG_{ct}$	0.0495	1	
$SNUCLEG_{ct}$	-0.3642	-0.3310	1

Table 2. Correlation matrix

In order to dispel any doubt we proceed as follows: i) we estimate the models excluding the variable volatility, concluding that there is no change in the coefficients' signals; ii) we compute the Variance Inflation Factor (VIF) test for multicollinearity (see table 3). The mean VIF is only 2.35 and the largest individual VIF is 4.21. From all this we conclude that collinearity is not a concern.

Variables	VIF	1/VIF
$SCOALEG_{ct}$	4.21	0.237790
$SNUCLEG_{ct}$	3.12	0.321027
$SGASEG_{ct}$	2.79	0.358631
$SOILEG_{ct}$	2.25	0.444951
$ENERGPC_{ct}$	1.98	0.504358
$LCRES_{ct-1}$	1.69	0.592946
$IMPTDP_{ct}$	1.65	0.604563
$VOLGDPPC_{ct}$	1.15	0.867271
Mean VIF	2.35	

Table 3. Variance Inflation Factor

Once the first inspection of the data had been made, we proceeded by testing the intrinsic characteristics of the data, namely by assessing the presence of the phenomena previously reported, i.e., heteroskedasticity, panel autocorrelation, and contemporaneous correlation. Table 4 reveals the specification tests we computed.

	Pooled	Random Effects	Fixed Effects
Modified Wald test (χ^2)			4885.68***
Wooldridge test $F(N(0,1))$	371.271***		
Pesaran's test		8.592***	8.069***
Frees' test		5.525***	5.749***
Friedman's test		62.200***	59.514***
Note: *** denotes 1% significance level.			

Table 4. Specification tests

From table 2, the null hypothesis of no first-order autocorrelation is rejected, as suggested by the Wooldridge test. From the Modified Wald statistic, we observe that the errors exhibit groupwise heteroskedasticity. As far as the contemporaneous correlation is concerned, all the tests are unanimous in their conclusions. They support the rejection of the null of cross-sectional independence, and thus the residuals do not appear to be spatially independent. The use of the PCSE is therefore sustained.

4. Results

After analysing the properties of the data, and since the pre-tests supported our choice for the estimations procedures, we proceeded to the presentation of estimation results, as well as their interpretation. Table 5 discloses the results and diagnostic tests.

		Dependent variable LGDP$_{ct}$			
Independent variables	OLS Model I	PCSE			
		Model II	Model III	Model IV	Model V
ENERGPC$_{ct}$	-0.0002*** (0.0000)	-0.0002*** (0.0000)	-0.0001*** (0.0000)	-0.0001*** (0.0000)	-0.0002*** (0.0000)
VOLGDPPC$_{ct}$	-0.7972*** (0.0412)	-0.7972*** (0.0436)	-0.4913*** (0.0571)	-0.4913*** (0.0676)	-0.4456*** (0.0630)
LCRES$_{ct-1}$	-0.0256*** (0.0676)	-0.2563*** (0.0316)	-0.0916** (0.0366)	-0.0916*** (0.0303)	-0.0920*** (0.0297)
IMPTDP$_{ct}$	-0.0086*** (0.0021)	-0.0086*** (0.0011)	-0.0028* (0.0015)	-0.0028** (0.0013)	-0.0059*** (0.0015)
SCOALEG$_{ct}$	-0.6137* (0.3599)	-0.6137*** (0.2032)	-0.2811 (0.2162)	-0.2811* (0.1678)	-0.3495** (0.1702)
SOILEG$_{ct}$	2.4772*** (0.7353)	2.4772*** (0.2998)	1.0848*** (0.3197)	1.0848*** (0.2359)	1.1918*** (0.2558)
SGASEG$_{ct}$	1.0171** (0.5107)	1.0171*** (0.3332)	0.4774* (0.2452)	0.4774** (0.1893)	0.6929*** (0.2012)
SNUCLEG$_{ct}$	2.2215*** (0.3674)	2.2215*** (0.1549)	1.3139*** (0.2601)	1.3139*** (0.1988)	1.4048*** (0.1855)
CONS	8.3756*** (0.4916)	8.3756*** (0.2644)	6.9737*** (0.2506)	6.9737*** (0.2556)	6.9991*** (0.2505)
Observations	376	376	376	376	376
R^2/Pseudo R^2	0.6465	0.6465	0.8555	0.8555	0.8961
F (N(0,1))	25.61***				
Wald (χ^2)		96981.67***	170.97***	656.20***	722.13***

Exclusion tests for VOLGDPPC$_{ct}$ and LCRES$_{ct-1}$

JST	188.35***	378.61***	76.59***	53.39***	52.11***
LRT	-1.0535*** (0.0834)	-1.0535*** (0.0559)	-0.5829*** (0.0709)	-0.5829*** (0.0825)	-0.5346*** (0.0759)

Exclusion tests for SCOALEG$_{ct}$, SOILEG$_{ct}$, SGASEG$_{ct}$, and SNUCLEG$_{ct}$

JST	32.11***	673.23***	51.07***	58.38***	70.24***
LRT	5.1021*** (1.5008)	5.1021*** (0.7610)	2.5949*** (0.6658)	2.5949*** (0.5056)	2.9401*** (0.5212)

Notes: OLS - Ordinary Least Squares. PCSE – Panel Corrected Standard Errors. The F-test is normally distributed N(0,1) and tests the null hypothesis of non-significance as a whole of the estimated parameters. The Wald test has χ^2 distribution. It tests the null hypothesis of non-significance of all coefficients of explanatory variables; JST - Joint Significance Test. JST is a Wald (χ^2) test with the null hypothesis of $H_O : \delta = \beta_k = 0$, with δ and β_k the coefficients of LCRES$_{ct-1}$ and the other explanatory variables, respectively. LRT - Linear Restriction Test has the null hypothesis of $H_O : \delta + \beta_k = 0$. All estimates were controlled to include the time effects, although not reported for simplicity. Standard errors are reported in brackets. ***, **, *, denote significance at 1, 5 and 10% significance levels, respectively.

Table 5. Results

Globally, results reveal great consistency and they are not dependent on the assumptions we made about variances across panels and serial correlations. There are no signal changes and, in general, the explanatory variables prove to be consistently statistically significant throughout the models.

The impact of both energy consumption *per capita* and import dependency on energy on economic growth is negative and statistically significant. The effect of the volatility on economic growth is negative and statistically highly significant. This result supports the assumption that higher volatility contributes to reducing economic growth. Results also provide strong evidence that the impact of energy on economic growth is dissimilar, varying according to the source of energy. While oil and nuclear reveal a positive and statistically highly significant effect on economic growth, it seems that renewables are hampering economic growth. This negative and statistically significant relationship is consistent throughout the several models. The effect of the fossil source natural gas on economic growth is positive and statistically significant, albeit at a lower level of significance (5% and 10%). This probably comes from the fact that this source is playing a recent role as a transition source from heavily polluting sources towards cleaner ones. The effect of coal on economic growth is not always statistically significant and, when significant, it is negative.

We deepen the adequacy of use of the variables $LCRES_{ct-1}$ and $VOLGDPPC_{ct}$ since their use is not widespread in the literature. Additionally, we test the simultaneous use of $SCOALEG_{ct}$, $SOILEG_{ct}$, $SGASEG_{ct}$, and $SNUCLEG_{ct}$. For that purpose, we provide two exclusion tests: i) Joint Significant Test - JST; and ii) Linear Restriction Test -LRT. The variables $LCRES_{ct-1}$ and $VOLGDPPC_{ct}$, together, must be retained as explanatory variables. Nevertheless, the sum of the estimated coefficients could not be statistically significant in explaining economic growth. From the LRT we reject the null hypothesis and then the sum of their coefficients is different from zero. The same conclusion is reached when we test the adequacy of the simultaneous control for the variables $SCOALEG_{ct}$, $SOILEG_{ct}$, $SGASEG_{ct}$, and $SNUCLEG_{ct}$. These variables must belong to the models. Together with the appropriateness of the use of PCSE, these tests corroborate the relevance of the explanatory variables, other than energy consumption per capita and import dependency on energy, since these are well described in the literature.

5. Energy consumption, dependency and volatility

To conclude that the higher the level of energy dependency, the lower the economic growth, is more intuitive than checking that the consumption of energy has the same negative impact on economic growth. However, looking carefully at these two relationships, both effects are understandable and expected. Regarding energy consumption, it is confirmed that the negative effect outweighs the positive one. As discussed above, this may be the result of two phenomena. On the one hand, this suggests that the additional consumption of energy stems from activities other than production, such as leisure activities. On the other hand, this additional consumption could be causing an overload in the external deficit of energy, for most EU Members.

The hypothesis that the dependency on energy imports is limiting economic growth is confirmed. Additional energy dependency means that the country becomes more subject to external constraints and to the rules, terms and prices set by other countries and external markets. Meanwhile, greater volume of energy imports is matched by financial outflows.

With respect to prices and diversification of primary energy sources, if larger energy dependency confers an advantage to the country, then it is likely that this dependency could have positive effects on economic growth. The reality is somewhat different, however. On the one hand, it appears that, in general, countries are price-takers in the international energy markets and, as such, they cannot influence prices. On the other hand, diversification of energy sources can lead to the need for diversified investments, which are expensive and are not sized to take advantage of economies of scale.

One of the common-sense ways to offset this negative effect will be the replacement of imports. To do so, countries can locally produce some of their energy needs, through the use of indigenous renewable resources. However, till now, the use of these resources to convert into electricity does not seem to produce the desired effects. On the contrary, it seems to limit the economic growth capacity of countries, in contrast to what happens with fossil energy sources.

Regarding the negative effect of volatility on economic growth, this result is in line with the hypothesis that the characteristic of irreversibility that is inherent in physical capital makes investment particularly susceptible to diverse kinds of risk (Bernanke, 1983; and Pindyck, 1991). Indeed, growth volatility produces risks regarding potential demand that hamper investment, generating a negative relationship between economic growth and its volatility. Other possible explanations are based on the learning-by-doing process, which contributes to human capital accumulation and improved productivity, which was assumed to be negatively influenced by volatility (e.g. Martin and Rogers, 2000).

6. Renewables *vs* traditional sources

By the end of the 21st century, it is accepted that we will no longer be using crude oil as a primary source of energy, as a consequence of its depletion. However, the coal situation is different. The reserves are large and will remain widely available for a long time, perhaps even for a century. Unfortunately, this source is both highly polluting and not so efficient. Similarly, natural gas will be available in larger quantities than the crude oil reserves, even considering that some of its reserves remain unknown. It will remain available as a primary source of energy even until the turn of the century. The conversion of natural resources into energy, mainly into electricity, is a matter of crucial importance within this context of changing the global energy paradigm.

With regard to the impact of different energy sources on economic growth, there seems to be a dichotomy between the effects that are caused by the use of renewable and traditional sources, which include fossil and nuclear sources. Both oil and natural gas stimulate economic growth in the period and countries considered, in line with what has been pointed out by the literature (e.g. Yoo, 2006) and with *the growth hypothesis*. The effect of coal on economic growth is statistically weaker than the other fossil fuels and, when statistically significant, this source of energy constrains economic growth.

Among the fossil fuels, oil is the source that has mostly contributed to economic growth. Given that the productive structures of the industrialised nations, such as those under review here, which are highly dependent on the intensive use of internal combustion engines, this effect was expected. Natural gas also has a positive effect on economic growth, although this source of energy has been particularly significant in recent years. This is due not only to the advances concerning the discovery of new reserves, but also to the considerable increase in the network of natural gas pipelines. At the same time, the combined cycle plants, which use mainly natural gas as fuel, have been used to guarantee electricity supply within the RE development strategy. This fact has contributed to stimulating the development of this energy source. It is a cleaner source, and is considered the transition source from fossil fuels to renewable sources.

Although the fact that RE limit economic growth is an unexpected result, it is one that deserves deep reflection in this chapter. Policy makers should be made aware of the global impacts of policies promoting the use of renewables. At first glance, the development of renewables should have everything to make it a resoundingly successful strategy. With this strategy, it would be possible to fight global warming, reduce energy dependency (not only economic but also geo-political), create sustainable jobs and develop a whole renewables cluster. What these results suggest is that the effects of renewables are more normative than real, i.e., the results are far from what they should be. Indeed, the development of renewables has been supported in public policies that substantially burden the final price of electricity available for final consumption to economic agents. At the same time, the productive structures of the countries are still heavily dependent on fossil-based technologies, such as internal combustion engines. Their conversion towards other technologies is a slow and expensive path.

7. The role that renewables play and what we want them to play

It is worth discussing, in more detail, the observed effect of renewables on economic growth. The main motivations for the use of RE are diverse, as indicated above. One of the most widely claimed is that of environmental concerns. Renewables allow traditional production technologies to be replaced with other cleaner technologies, with lower emissions of greenhouse gases, in line with what is suggested by De Fillipi & Scarano (2010). The question that many countries, such as the United States of America, have raised is that this substitution severely limits the capacity for growth. This is the ultimate cause for the non-ratification of important international treaties like the Kyoto Protocol.

Moreover, it is far from unequivocally proven that more intensive use of renewables contributes decisively to the reduction of CO_2 emissions, in line with what was pointed out, for example, by Apergis et al. (2010). In this chapter we tested the inclusion of CO_2 emissions as an explanatory variable, but it proved not to be statistically significant.

Renewable sources should be placed within the mix of energy sources, requiring the simultaneous use of other sources, mostly fossil. The intermittency of renewables cannot be compensated by the use of nuclear energy. The offset of the lack of production from renewables implies the ability to frequently turn these other sources of support on and off, which is obviously not possible when it comes to nuclear energy. The counterbalance has to be made by fossil fuels, mainly natural gas and coal. The latter is a cheaper source of energy but at the same time is also highly polluting.

The growing use of RE has been heavily dependent on policy guidance. Most EU Members, either voluntarily or compulsorily, have established several mechanisms to support these alternative sources of energy. One of the most commonly used policies is the feed-in tariff, which consists of setting a special price that rewards energy from clean sources. This policy and all other public policies lead to government expenses. These costs are passed on by the regulators to the final consumer, both residential and firm consumers. When they are not passed on by regulators in the regulated market, then in the liberalised market, the producers transfer to consumers the extra costs they have when producing energy from renewable sources. This strategy of promoting RE can thus burden the economy with electricity costs that are too high and therefore hinder economic growth.

It is already clear that the overall strategy for electrification of the economy requires large volumes of financial resources, which may be diverted from other alternative projects. However, the massive investment in renewables may promote divestments, not only in the technological upgrades of other conventional sources of energy, but also in other industrial projects. In order to be able to achieve compliance with the requirements of market entry, and to keep innovating mainly through R&D, players in renewables are obliged to issue debt. Given that the available financial resources are scarce, this debt from renewables may be preventing players in other industries, with even greater multiplier effects on economic growth, from achieving fair interest rates which do not compromise the appropriate return.

In this regard, it is worth highlighting that another factor which may help explain the negative effect of renewables on economic growth is that the investment should be paid during its usable life, as good practices suggest. The reality shows that this normally does not happen. Consumers have to start to bear the cost of a wind farm or solar park almost immediately. More serious still is that the Government requires the payment for a licence allocation of power generation in advance. After that, the Government guarantees prices for the purchase of the electricity generated. Finally, the winners of the bids will capture the regulators to immediately recover these costs of entry. Overall, this has little to do with the nature of renewable technology. Instead, it is more a launch of a tax resulting from renewables diverted to electricity costs and, ultimately, on consumers and on the economy as a whole.

No less important within this discussion of the effect of renewables on economic growth is the effect brought about by renewables on the technology and production capacity already installed. In fact, greater use of renewables implies the dismantling or simply the creation of excess capacity based on conventional sources. Note that, in the past, these sources represented a major cause of the degree of development, industrialisation and prosperity of the countries. They grew mainly supported on technologies based on fossil sources of energy. The increased use of renewables can thus be causing two outcomes. On the one hand, renewables may diminish the positive effect of conventional sources on economic growth. On the other hand, renewables discourage technological upgrades of conventional sources. Nonetheless, these sources can still evolve, both as regards the level of energy efficiency (thus reducing dependency), and as regards greenhouse gas emissions.

It should be noted that the results presented in this chapter were obtained by studying evidence from 1990 to 2007. They do not allow us to unequivocally conclude that RE will not stimulate economic growth in the near future. Indeed, the studies using official statistics on energy produced from renewable sources inevitably suffer from a problem that could lead to some kind of bias in the results. The official statistics on the use of RE normally do not reveal the true contribution of these sources both to our lives and to the economy as a whole. There is a plethora of examples that illustrate this failure in the statistics. When the sun comes through the windows of our homes or businesses, effectively heating them, there are significant energy savings by avoiding the use of traditional energy sources. The statistics also do not capture the effect when the sun heats the water that we use both for bathing and industrial activities. The sunlight that enables the achievement of sporting events, entertainment and various economic activities without resorting to light bulbs is a valuable contribution of this renewable source, but it is also absent from official statistics.

Solar radiation allows the growth of plants, both for biomass and food, which in turn creates energy. Finally, it should not be forgotten that solar radiation allows the chemical process for the formation of fossil fuels. The natural resource water does not only provide the water supply for dams for electricity generation, with the particularity of this feature in allowing storage. In short, by not considering all these effects from renewables, the results that come from the use of official sources of statistics may not give the full picture of the effect of renewables. All the energy that results from natural and renewable sources is generally not included in the statistics, but it is an invaluable contribution to reducing the use of other sources, mainly polluting fossil sources.

In general, if taken together, renewables are likely to contribute positively to the process of economic growth. However, regarding the use of natural sources for electricity generation through direct human intervention, such as wind and photovoltaic facilities, it seems that the desired results are still a long way off. In fact, this may distort the conclusions about the contribution of renewables to economic growth. The immediate challenge will therefore be to strengthen the use of these renewable sources, in their natural state. In other words, both the organisation of society and the economy should be more consistent with the maximisation of benefits from these natural sources. Just two simple examples. First, more energy-efficient houses must be built. They should maximise the benefits of solar power for heating, while wind, rain and vegetation should contribute to cooling them. Second, both sports and musical shows should be performed during periods when natural light eliminates the need for artificial lighting, which consumes a great deal of electricity.

Overall, a country's decision to intensify the use of the RE mix is eminently political, rather than economic. In this process, there are two strongly related factors that will influence the role of renewables in the economy. The first concerns the evolution of technology converting energy emitted by renewable sources into usable energy, such as electricity. The second factor is of a political nature. The consequences for renewables will be rooted in this political process. We believe it is essential that the regulatory authorities do not excessively and quickly pass costs of RE production to the economy. Instead, they should commit players operating in this industry to assuming a significant part of the risks inherent in these energies.

8. Conclusion

This chapter is centred round the interaction between economic growth and its main drivers, focusing mainly on the effect of each energy source, distinguishing between traditional sources and renewables. We go on to shed some light on the relevance of developing the use of renewables in the energy mix and on their consequences in relation to economic growth. To do so, we apply panel data techniques to a set of EU 21 Members, for the time span 1990-2007. Overall, the results prove to be consistent and the use of the Panel Corrected Standard Errors estimator seems to be suitable, matching the data properties.

Both energy dependency and volatility have contributed negatively to economic growth. Conventional wisdom indicates that the use of energy generated from renewable sources can contribute both to reducing this dependency and to reducing volatility. Renewable energy is produced locally and thus contributes to energy self-sufficiency. Meanwhile, the contracts for generation from renewables are generally medium to long term, which are characterised by lower uncertainty as to price behaviour. The results suggest, however, that renewables are also hampering economic growth, in the period and for the countries analysed. This chapter discusses extensively the possible reasons for this effect caused by renewables. It is confirmed that the traditional sources of energy have been real engines of economic growth, although the role played by each of these sources is not homogeneous. Among the fossil fuels, oil has played a key role in the process of economic growth.

On a daily basis, we use renewables without noticing. Accordingly, we directly make use of renewable energy in its natural state, such as it is available on Earth, like in water heating, lighting or heating our homes. This generous contribution from nature, however, is usually absent from the statistics. With regard to the use of technology for conversion of renewable energy into usable energy, mainly from sun and wind, the conclusions are dissimilar. Using the statistics, we find that the share of renewables in total energy supply is not having the desired effect, as far as economic growth and wealth creation are concerned. Ultimately, with the current state of affairs, the decision to invest in renewable energy remains essentially political.

9. Acknowledgement

We gratefully acknowledge the generous financial support of the NECE - Research Unit in Business Science and Economics, sponsored by the Portuguese Foundation for the Development of Science and Technology.

10. References

Akinlo, A. (2008). Energy consumption and economic growth: evidence from 11 Sub-Sahara African countries. *Energy Economics*, 30, pp. 2391-400.

Apergis, N. & Payne, J. E. (2010). Renewable energy consumption and economic growth: Evidence from a panel of OECD countries. *Energy Policy*, 38, pp. 656-60.

Apergis, N. Payne, J.E., Menyah, K. & Wolde-Rufael, Y. (2010). On the causal dynamics between emissions, nuclear energy, renewable energy, and economic growth. *Ecological Economics*, 69, pp. 2250-60.

Baum, C. (2001). Residual diagnostics for cross-section time series regression models. *The Stata Journal*, 1, pp. 101-4.

Bean, C. R. (1990). Endogenous growth and the pro-cyclical behaviour of productivity. *European Economic Review* 34, pp. 355-94.

Bernanke, B. S. (1983). Irreversibility, uncertainty, and cyclical investment. *Quarterly Journal of Economics*, 98, pp. 85-106.

Black, F. (1987). *Business cycles and equilibrium*. New York: Basil Blackwell.

Blackburn, K., & Pelloni, A. (2004). On the relationship between growth and volatility. *Economics Letters*, 83, pp. 123-7.

Blackburn, K. (1999). Can stabilization policy reduce long-run growth? *Economic Journal*, 109, pp. 67-77.

Cameron, A. & Triverdi, P. (2009). *Microeconometrics using Stata*, Stata Press. StataCorp, Texas.

Chiou-Wei, S. Chen, C-F. & Zhu, Z. (2008). Economic growth and energy consumption revisited - evidence from linear and nonlinear Granger causality. *Energy Economics,* 30, pp. 3063-76.

De Filippis, F. & Scarano, G. (2010). The Kyoto Protocol and European energy policy. *European View,* 9, pp. 39-46.

EU. (2001). Directive 2001/77/EC on the promotion of electricity produced from renewable energy sources in the internal electricity market.

EU. (2009). Directive 2009/28/EC on the promotion of the use of energy from renewable sources and amending and subsequently repealing Directives 2001/77/EC and 2003/30/EC.

Fang, W-S. & Miller, S. M. (2008). The Great Moderation and the Relationship between Output Growth and its Volatility. *Southern Economic Journal,* 74, pp. 819–38.

Fountas, S., Karanasos, M., 2006. The relationship between economic growth and real uncertainty in the G3. *Economic Modelling,* 23, 638–647.

Frees, E. (1995). Assessing Cross-Sectional Correlation in Panel Data. *Journal of Econometrics,* 69, pp. 393–414.

Frees, E. (2004). Longitudinal and Panel Data: Analysis and Applications in the Social Sciences. *Cambridge University Press.*

Friedman, M. (1937). The Use of Ranks to Avoid the Assumption of Normality Implicit in the Analysis of Variance. *Journal of the American Statistical Association,* 32, 675-701.

Friedman, M. (1968). The role of monetary policy. *American Economic Review,* 58, 1–17.

Frondel, M. Ritter, N. Schmidt, C. & Vance, C. (2010). Economic impacts from the promotion of renewable energy technologies: The German experience. *Energy Policy,* 38, pp. 4048-56.

Grier, K. B., & Tullock, G. (1989). An empirical analysis of cross-national economic growth, 1951–80. *Journal of Monetary Economics,* 24, pp. 259–76.

Hodrick, R., & Prescott, E.P. (1997), "Postwar Business Cycles: An Empirical Investigation. Journal of Money, Credit, and Banking, 29, pp. 1–16.

Huang, M-Y., Alavalapati, J., Carter, D. & Langholtz, M., (2007). Is the choice of renewable portfolio standards random? *Energy Policy,* 35, pp. 5571-5.

Kneller, R., & Young, G. (2001). Business cycle volatility, uncertainty and long-run growth. *Manchester School,* 69, pp. 534–52.

Kormendi, R., & Meguire, P.G. (1985). Macroeconomic determinants of growth: Cross-country evidence. *Journal of Monetary Economics,* 16, 141–63.

Lee, C-C. & Chang, C-P. (2007). The impact of energy consumption on economic growth: Evidence from linear and nonlinear models in Taiwan. *Energy,* 32, pp. 2282-94.

Lucas, R. E. (1972). Expectations and the neutrality of money. *Journal of Economic Theory,* 4, pp. 103–24.

Marques, A.C., Fuinhas, J.A. & Manso, J. (2011). A Quantile Approach to Identify Factors Promoting Renewable Energy in European Countries. *Environmental and Resources Economics,* 49, pp. 351-66.

Martin, P., & Rogers, C. A. (1997). Stabilization policy, learning by doing, and economic growth. *Oxford Economic Papers*, 49, pp. 152–66.

Martin, P., & Rogers, C. A. (2000). Long-term growth and short-term economic instability. *European Economic Review*, 44, pp. 359–81.

Menegaki, A. N. (2011). Growth and renewable energy in Europe: A random effect model with evidence for neutrality hypothesis. *Energy Economics*, 33, pp. 257-263.

Menyah, P.K. & Wolde-Rufael, Y. (2010). CO2 emissions, nuclear energy, renewable energy and economic growth in the US. *Energy Policy*, 38, pp. 2911-15.

Miller, S. M. (1996). A note on cross-country growth regressions. *Applied Economics*, 28, pp. 1019–26.

Mirman, L. (1971). Uncertainty and optimal consumption decisions. *Econometrica*, 39, 179–85.

Narayan, P. K. & Smyth, R. (2008). Energy consumption and real GDP in G7 countries: New evidence from panel cointegration with structural breaks. *Energy Economics* 30, pp. 2331-41.

Odhiambo, N. (2010). Energy consumption, prices and economic growth in three SSA countries: A comparative study. *Energy Policy*, 38, pp. 2463-69.

Ozturk, I. (2010). A literature survey on energy–growth nexus. *Energy Policy*, 38, pp. 340-9.

Payne, J.E. (2010). Survey of the international evidence on the causal relationship between energy consumption and growth. *Journal of Economic Studies*, 37, pp. 53-95.

Pesaran, M. H. (2004). General diagnostic tests for cross section dependence in panels. *Cambridge Working Papers in Economics*. Faculty of Economics, University of Cambridge, 0435.

Phelps, E. S. (1968). Money wage dynamics and labor market equilibrium. *Journal of Political Economy*, 76, pp. 678–711.

Pindyck, R. S. (1991). Irreversibility, uncertainty, and investment. *Journal of Economic Literature*, 29, pp. 1110–48.

Rafferty, M. (2005). The effects of expected and unexpected volatility on long-run growth: Evidence from 18 developed economies. *Southern Economic Journal*, 71, pp. 582–91.

Ramey, G. & Ramey, V.A. (1995). Cross-country evidence on the link between volatility and growth. *American Economic Review*, 85, pp. 1138–51.

Reed, W.R. & Ye, H. (2009). Which panel data estimator should I use? *Applied Economics*, 43, pp. 985-1000.

Saint-Paul, G. (1993). Productivity growth and the structure of the business cycle. *European Economic Review*, 37, pp. 861–90.

Toklu, E., Güney, M.S., Isık, M., Comaklı, O., & Kaygusuz, K. (2010). Energy production, consumption, policies and recent developments in Turkey. *Renewable and Sustainable Energy Reviews*, 14, pp. 1172–86.

Vachon, S. & Menz, F. (2006). The role of social, political, and economic interests in promoting state green electricity policies. *Environmental Science & Policy*, 9, pp. 652-62.

Wolde-Rufael, Y. (2009). Energy consumption and economic growth: the experience of African countries revisited. *Energy Economics*, 31, pp. 217-24.

Yoo, S-H. (2006). Oil Consumption and Economic Growth: Evidence from Korea. *Energy Sources, Part B: Economics, planning and policy*, 1, pp. 235-43.

2

A Review on the Renewable Energy Resources for Rural Application in Tanzania

Mashauri Adam Kusekwa

Electrical Engineering Department, Dar es Salaam Institute of Technology, Dar es Salaam, Tanzania

1. Introduction

United Republic of Tanzania with a surface area of about 945,087 square kilometres is located in East Africa bordering the Indian Ocean to the East, Mozambique, Malawi and Zambia to the South, The Democratic Republic of Congo (DRC), Rwanda and Burundi to the West and Uganda and Kenya to the North. The country has a population of 41,915,799 [Economic survey, 2009], of which 21,311,150, equivalent to 50.8 percent are female, while 20,604,730 equivalents to 49.2 percent are male. Tanzania mainland has a population of 40,683,294, while Tanzania Zanzibar has a population of 1,232,505. Population distribution shows that 31,143,439 people, equivalent to 74.3 percent of the population live in rural areas, while 10,772,360 people live in urban areas. The population is based on the growth rate of 2.9 percent per annum estimated during the 2002 population and housing census of 2002.

Geographically, the country lies between Latitudes 1-12⁰S and Longitude 20-41⁰. Climatically, the country is tropical, hot with arid central plateau surrounded by Lake Victoria in the North-West, Lake Tanganyika in the West, temperate highlands in the North and South; the Coastal plain facing the Indian Ocean and Mount Kilimanjaro being the highest mountain in Africa. About 62,000 square kilometers [Casmiri, 2009] of the land is covered by water, including the three fresh-trans-boundary lakes of Victoria, Tanganyika and Nyasa. Woodlands account for more than 33,500 square kilometers and arable land; land suitable for agriculture is concentrated in Southern, North-East and Central part of the Country, covering more than 44 million hectares.

1.1 Background

Energy is an essential factor in both livelihoods and industrial development. An increase in unsuitable use and excessive consumption of energy has been causing not only local pollution but also global environmental problems such as global warming and climate change. In addition; fossil energy sources such as coal, oil, etc are so limited, if energy security is not fully ensured, such unsuitable use may pose a significant threat to economic activities and even to people's lives. Therefore, to realize sustainable development, a stable energy supply is important as well as an improvement in energy related environmental problem. Renewable energy resources and technologies are one of kind of approaches that if well utilized can play part in enhancing energy security and avoidance of environmental problems emanating from fossil energy sources.

Renewable is a term used for forms of energy which are not exhaustible by use over time. It means that the renewable sources can be regenerated or renewed in a relatively short time. This section focuses on leading renewable energy sources in Tanzania. The leading sources are a result of the assessment conducted in the country. The following are leading renewable sources: biomass, solar, hydropower, wind, solar and geothermal. However, hydropower as one of the leading renewable energy sources in the country is not discussed in this chapter.

In general, the sources of renewable energy can be divided, according to their origin, into natural renewable sources i.e. wind, solar, geothermal etc., and renewable sources resulting from human activity which include: biomass including landfill gas and industrial heat recovery.

The energy balance of the country shows that biomass use accounts for 88 percent of energy consumption in particular in the rural area [Magesa, 2007]. The majority of the rural population relies on biomass as fuel for cooking. Biomass is followed by Petroleum (7 percent), gas (2 percent) etc. Summary of the primary sources in the country is given in Table 1.

Petroleum exploration efforts have been made in the past and are still going on but so far no oil has been found. Therefore, the country relies exclusively on imports of its oil whereby the transport sector consumes more than 40 percent of all petroleum imported. At present, only 70 percent of the demand for petroleum fuel is met. The increase in importation of petroleum products and continuous rise in oil price is heavy burden for the country. With the introduction of right policy, regulation and incentives, the country has a potential of substituting a large percentage of the imported fuel with Biofuel that could be produced within the country.

Source of Energy	Composition in percentage[%]
Petroleum	7
Electricity	1.4
Renewable (Solar, Wind, etc)	1.3
Biomass	88
Others (e.g. Coal)	0.3
Gas	2
Total	100

Table 1. Primary Energy Sources (April 2011)

Renewable energy can effectively solve the problem of global warming and climate change being experienced in the country; in addition, renewable energy technologies can create jobs to young graduate and hence reduce poverty. Promotion of renewable energy technologies has not progressed easily in the country due to economic inefficiency when competing with traditional energies of oil and natural gas. There are two premises from which to promote renewable energy technologies on a large scale use. First, is diversification of the risk to the environment and social-economic activities; secondly, to increase stable energy supply and enhance energy security of the country. Use of renewable energy technologies in the country will have a positive impact on social-economic development of the country in the future.

This chapter focuses on assessment of renewable energy technologies as an alternative approach in electrifying rural Tanzania. In rural Tanzania, there is still an excessive demand and dependence upon traditional energy use. Developing appropriate technologies, efficient extraction of energy from renewable energy sources and use of modern renewable energy technologies to store the generated energy in more efficient manner that has a significant potential to mitigate climate change, offer a sustainable energy supply, create jobs, reduce poverty and achieve a sustainable development.

2. Available energy resources in Tanzania

A large portion of the United Republic of Tanzania remains un-electrified. The vast part of the country has vast reserves of natural energy resources including water, natural gas, coal, wind, solar, ocean waves, uranium and even geothermal energy. If these resources could be harnessed they could meet the ever-growing demand of electricity for many years to come and create opportunities to export electricity to neighbouring countries. In the following section, available energy resources in Tanzania is presented

2.1 Hydropower
Hydropower currently contributes more than 50 % of electricity generated in the country. Given that supply is not meeting demand; deliberate efforts are needed to be taken to look for other sources. Large areas of the country are supplied with power from hydro stations which include Hale, Kidatu, Kihansi, New pangani Falls, Mtera and Nyumba ya Mungu. Large reservoirs are located at Mtera, Kidatu and Nyumba ya Mungu with storage Capacity of about 4,200 Million cubic metres [Casmiri, 2009] while Hale, Pangani fall and Kihansi have three head ponds with a total capacity of 2.26 Million cubic metres. Electricity generated from hydropower is given in Table 2. The re-filing of the above mentioned reservoirs depends on the availability of sufficient rainfall from various basin including Rufiji, Ihefu and Pangani basins. Therefore, the contribution of hydropower to the energy mix of the country varies according to climatic conditions.

Energy Source	Plant Name	Installed Capacity [MW]
Hydropower	Kidatu	204
Hydropower	Kihansi	180
Hydropower	Mtera	80
Hydropower	New Pangani Falls	68
Hydropower	Hale	21
Hydropower	Nyumba ya Mungu	8
TOTAL		561

Table 2. Electricity Generated from Hydro Source (Source TANESCO, 2009)

Future hydropower projects under plan are given in Table 3
Hydropower resources currently contribute more than 50% of electricity generated in the country. It is a leading renewable energy resource. However, electricity generated from hydropower highly depends on weather conditions. Due to climate change being experienced in the country, it has been observed that the pattern of rain in the catchment

area is not consistent; hence the level of water in all the dams used to generate electricity is falling. This trend has affected distribution of electricity in the country. TANESCO, the Government owned utility company responsible for transmission and distribution of electricity has resorted to introducing load shedding.

Hydropower is a leading renewable energy resource in the country but it cannot guarantee sustainable supply of electricity. Other resources are needed to supplement energy during drought. Renewable resources such as biomass, wind, solar, geothermal, etc., must be exploited as supplement to hydropower resources.

2.2 Natural gas

The country has abundant natural gas reserves in the coastal basin that are estimated at more than 45 billion cubic metres [TIC, 2007]. Significant gas discoveries have been made on the coastal shores of Indian Ocean. Four discoveries of natural gas fields so far have been established in the vicinities of Songo Songo Island (about 250 km south of Dar es salaam in 1974), Mukuranga (about 60 km South of Dar es Salaam, in December 2007), Mnazi Bay (about 450 km south of Dar es Salaam in 1982) and Kiliwani North (about 2.5 km South East of Songo Songo Island in April, 2008), but only two gasfield i.e. Songo Songo and Mnazi Bay are producing.

Songo Songo gasfield was estimated at 810 billion standard cubic feet, while proven, probable and possible reserves stood at 1.1 trillion standard cubic feet. Mnazi bay gas reserves are estimated at 2.2 trillion standard cubic feet. The gas from Songo Songo Island is transported by pipeline to Dar es Salaam where it is distributed to electricity generation plant and industries especially cement industry. Natural gas supplied to Songas Power Plant generates about 200 MW of electricity. The generated electricity is fed into the National Electricity grid and distributed to end users by TANESCO. In 2010, TANESCO started operating its own plant at Ubungo to generate 102 MW from natural gas. Symbion Power Plant is a private company using natural gas to generate 112.5 MW. Tegeta is generating 45 MW from natural gas. It is anticipated that in the near future more IPPs will generate power using the same gas from Songo Songo gas field.

Energy Source	Plant Name	Installed Capacity [MW]
Hydropower	Stiegler's Gorge	2,100
Hydropower	Mpanga	165
Hydropower	Ruhudji	358
Hydropower	Rumakali	222
Hydropower	Lukose & Masigira	118
Hydropower	Rusumo Falls	21
TOTAL		**2,984**

Table 3. Future Hydropower projects (Source EWURA Annual Report 2008/09)

In Mtwara the same gas is extracted at Mnazi bay and is used to generate electricity to Mtwara and Lindi Regions. The two regions are not connected to the National electricity grid. Electricity generation from natural gas is increasing and it is anticipated that in the near future, natural gas will replace hydropower as source of electricity generation in Tanzania. Electricity generated from natural gas is given in Table 4.

Energy Source	Plant Name	Installed Capacity [MW]
Natural gas	Songas	200
Natural gas	TANESCO	103
Natural gas	Symbion	112.5
Natural Gas	Tegeta	45
TOTAL		**460.5**

Table 4. Electricity Generated from Natural gas Source (Source TANESCO, 2011)

Future thermal power projects under plan are given in Table 5

Energy Source	Plant Name	Installed Capacity [MW]
Natural gas	Kinyerezi	240
Coal	Kiwira 1	200
Coal	Mnazi Bay	300
Coal	Mchuchuma	400
Coal	Ngaka	400
Natural gas	Dar es Salaam	100
HFO	Nyakato	60
TOTAL		**1,700**

Table 5. Future Thermal Power generation from coal and natural gas **(Source MEM 2011).**

2.3 Coal

Coal is another resource of primary fuel available in the Country. Coal is found in Kiwira, Mchuchuma and recently in Ngaka. It is estimated that the country has more than 1,200 Million metric tons of coal. Kiwira coal mine supplied between 4-6MW of electricity to the grid annually when it was working; However, now it is closed. Current plan is to revamp the mine so that it can be able to generate power so as to curb power shortages especially during drought seasons when hydro plants are affected. About 1.5 million tons per annum are expected to be mined at the Mchuchuma coalfield where about 400 MW thermal power plants will be built in an effort to increase reliability and security of grid power. Ngaka mine is also expected to be functional in the near future. Apart from grid electricity generation, coal is used in some industries such as cement and paper mills. However, its use is still low. In this aspect its contribution to energy mix of the country is almost negligible.

2.4 Petroleum

The consumption of petroleum and related products in Tanzania is about 1.54 million cubic metres annually [Casmiri, 2009]. Petroleum is imported from the Persian Gulf and the Mediterranean region. Most of petroleum depots are in Dar es Salaam near the Dar es Salaam harbour. From Dar es Salaam petroleum is transported to up-country regions via trunk roads and some areas by Tanzania Railways. Therefore, for the country to have access to petroleum products, infrastructures such as trunk roads, railways should be passable throughout the year regardless of climatic conditions.

Imported petroleum and related products are widely used in the transport and industrial sectors. It is also used for generating electricity in isolated grid-diesel power stations that have an installed capacity of about 21 MW and are located in Songea, Masasi, Tunduru, Kilwa Masoko, Mpanda, Kigoma, Biharamulo, Ikwiriri, Mafia and Ngara. Petroleum and related by-products are imported and distributed by private companies regulated by the Energy and Water Utilities Regulatory Authority (EWURA) which has the authority of monitoring performance and standards with regards to quality, health, safety and environment, licensing, tariff review of electricity and price control.

The current level of energy demand and supply in the country signifies low level development in the industry sector, transport, and commerce. Industry and urban households depend to a considerable extent on energy sources such as electricity and petroleum products which are either imported (petroleum) or generated in the country (electricity). Traditional segment of the economy, mainly rural households depend on biomass as the main source of energy. Semi-urban and urban dwellers also depend on biomass especially charcoal and firewood as a source of energy for cooking purposes despite the fact that a large number of households in this category have access to electricity.

The demand for modern energy i.e. electricity is growing at a fast rate. From 1990-1998 demand for electricity rose by 4.45 %; from 2003-2006 demand rose by 8% despite a prolonged period of electrical power shedding due to drought and insufficient rainfall for hydropower catchments areas. The demand for electricity is expected to increase from the present value of 925 MW to at least 3,800 MW by 2025 [Msaki, 2006]. Despite low electricity consumption estimated at 14 % for urban areas and about 2% for rural areas, in general, supply is still unable to meet demand. This shortfall is attributed to the country's dependence on hydropower which in turn is affected by climate variation and climate change. To increase accessibility of electricity to both urban and rural areas necessary efforts are needed. In this aspect, the government decided to commit itself to facilitate the increase of use of renewable energy as an alternative solution for increasing accessibility of modern energy to rural areas. Therefore, a number of reforms i.e. legal framework measures, policies and strategies have been formulated and enacted to provide a constructive atmosphere for utilization of renewable energy resources in the country. The following are some of the policies and strategies adopted for the promotion and facilitation of an increased use of renewable energy within the country.

3. Legal framework and policies

Tanzania power sector has undergone through different turbulent periods, changes and reforms since the country attained its independence in 1961. Most of the changes and reforms have sent positive signals to those who are interested in developing or starting electricity project in the country. The changes and reforms include laying down a National Energy policy (NEP), Electricity Industry Policy and The electricity act of 2008, and guidelines for sustainable liquid biofuel development. Some of the legal framework and policies are elaborated in the following section.

3.1 National energy policy (NEP) -2003
The first national energy policy (NEP) for the Country was formulated in 1992. Since then the energy sector has undergone a number of changes, necessitating adjustments to the

initial policy. These changes include change in the way the role of the government from service provider to service facilitator, liberalization of the market and encouragement of private sector investment. With these changes, the energy policy of 1992 was replaced in 2003.

The objective of the 2003 NEP is to ensure availability of reliable and affordable energy supply and use in a rational and sustainable manner in order to support national development goals. The National Energy Policy of 2003 aims to establish energy production, procurement, transmission, distribution and end-user systems in an efficient, environmentally sound, sustainable and gender-sensitized manner.

Key objectives of the 2003 NEP regarding to Renewable Technologies (RT) and services include:

- Encourage efficient use of alternative energy sources.
- Facilitate Research and Development (R&D) and application of Renewable Energy for electricity generation.
- Facilitate increased availability of energy service including off-grid electrification of rural areas.
- Introduce and support appropriate fiscal, legal and financial incentives for Renewable Energy Technologies.
- Ensure the inclusion of environmental consideration in energy planning and implementation.
- Support Research and Development (R&D) in Renewable Energy Technologies.
- Establish norms, codes, of practice, standards and guidelines for cost-effective rural energy supplies and for facilitating the creation of an enabling environment for the sustainable development of renewable energy sources.
- Facilitate the creation of an enabling environment for sustainable development of Renewable Energy Sources.
- Promote entrepreneurship and private initiatives for the production and marketing of products and services for rural and renewable energy.
- Ensure priority on power generation capacity based on indigenous resources.

The policy encourages public and private partnerships to invest in the provision of energy services. It also seeks to promote private initiatives at all levels and stresses the need to make local and foreign investors aware of the potential of the Tanzanian energy sector. To implement the policy several laws have been enacted, among them are:

3.2 Rural energy act (2005)

The Rural Energy Agency (REA) and the Rural Energy Fund (REF) are autonomous bodies established under the Rural Energy Act no. 8 of 2005. The two bodies are monitored by the Ministry of Energy and Minerals (MEM). REA and REF are established to:

- Promote, stimulate, facilitate and improve energy access for social and commercial use in rural Tanzania.
- Promote the rational and efficient generation and use of energy.
- Utilize the rural energy fund (REF) to finance suitable rural energy projects
- Facilitate activities of key stakeholders with interest in generation and electrification of rural areas.
- Provide capital subsidies to rural energy projects through a trust fund
- Utilize the REF to finance viable rural energy projects.

- Allocate resources to projects in open and transparent manner and with well defined allocation criteria.

The act provides REF with funds from the following sources.

- Government budgetary allocation.
- Contribution from international financial organizations and other development partners.
- Levies of up to 5% on the commercial generations of electricity from the national grid
- Levies of up to 5% on the generation of electricity in specified isolated systems
- Fees for programmes, publications, seminars, consultancy activities and other services provided by the agency
- Interests or returns on investment.

REA/REF have already supported various off-grid projects in small hydro power projects, biomass cogeneration projects, biomass gasification projects in Mafia and Mkonge Energy project. The supported projects are currently at various stages of implementation. The total expected capacity is 46 MW. A total of 8,400 new connections are expected. REA/REF support fiscal incentives for rural energy projects and programmes and count amongst the National aid initiatives attracting fiscal initiatives. On top of government subsidy to REF, the agency is also allowed to take up to 5% surcharge on each unit of energy generated by commercial electricity producer. REA/REF subsidies also support solar PV Systems. However, the subsidy is limited to 100Wp for domestic use and up to 300Wp for Institutions.

3.3 Electricity act (2008)

The electricity Act of 2008 replaces the electricity ordinance Cap 131 of 1931. The act implements the National Energy Policy of 2003. The act opens up the electricity sector for generation, transmission, distribution and sales to private sector participation. It provides instruments for the regulator (EWURA) and stipulates the roles of Rural Energy Agency (REA) and Rural Energy Fund (REF) and sets the general conditions for cost effective tariffs and least-cost electrification options in particular to rural areas.

In addition, the act recognizes other activities such as:

- The preparation of rural electrification strategies
- Plan to promote access to electricity in rural Tanzania
- Recognize Fair Competition Commission
- Standardized small power purchase agreement for 100kW to 10 MW
- Standardized Power Purchase Tariffs and Fair Competition Tribunal
- Power Sector Master Plan (PSMP) to be updated annually.
- Electricity to be generated from any primary source including renewable energy.

The electricity act of 2008 has opened up windows for renewable energy promotion in particular in rural areas. It is anticipated that more research and development in renewable energy would increase competitiveness of renewable energy technologies.

3.4 Guidelines for sustainable liquid biofuels development (2009)

Advanced technologies available today facilitate production of liquid biofuels and generation of electricity using solid by-product through cogeneration. Using current technologies, economically it is feasible to produce biofuels through the use of agriculture crops which in some cases are also food crops. In this aspect, the government of Tanzania is aware of potential benefits that could be realized through development of the biofuels industry; these include technology transfer through new bio-energy industries, employment and income generation in industry and agriculture sectors, improved energy security, foreign exchange savings via the reduction of oil import, increased foreign exchange through export of biofuels and reduced emission of pollutants and other harmful particles.

In order to create an avenue for biofuel development, the government has published guidelines for sustainable liquid biofuels development, which include:

- Application and Registration procedures for biofuels investments
- Permit and fees
- Taxation and incentives
- Land Acquisition and use
- Contract farming
- Sustainability of biofuel production
- Farming approaches and seed management
- Efficient utilization of biofuel crops
- Appropriate infrastructure development
- Community engagement
- Processing of biofuels
- Storage and handling of biofuels
- Transportation and distribution
- Quality of biofuels (quality standard)
- Blending (biofuel and mineral fuel)
- Biofuel waste management (use, re-use, recycling and disposal)
- Research and Development (condition to fund or support research and development)

The guidelines will attract more investors to come and invest in the country. It is anticipated that in the near future, biofuels will contribute massively to the energy mix of the country.

Modern energy services require the growing inclusion of renewable energy into the sustainable energy mix of the country. The legal frameworks and policies have already been enacted and are in place. The task ahead is how to implement. However, this task is not easy; it needs concerted efforts, organisation and proper planning which include identifying the leading renewable energy resources in the country. A brief summary of leading renewable energy resources is presented in section 4.

4. Leading renewable energy resources in Tanzania

Although biomass is the main source of energy in Tanzania particularly in the rural area, the country is still relying heavily on imported commercial energy in the form of oil and petroleum products; characteristic of all non-oil producing economies. In this aspect, most planners have simplified their work by directing their attention on fossil-fuel, especially petroleum where data is easily available. Thus, more investigation has been on commercial fuels and less on biomass fuel or other renewable energy sources. However, as the effect of fossil fuel on the environment and climate change is becoming serious than before, the attention is now shifting towards renewable energy resources utilization. As this shift is taking pace, more research and resources must be undertaken and used in developing renewable energy technologies for sustainability of the country. In this sub-section the focus is on establishing leading renewable energy sources in the country which can be used as input in renewable energy technologies in generation of energy.

Modern biomass comprises a range of products derived from photosynthesis and is in fact chemical solar energy storage in nature. This type of renewable energy represents a renewable storage of carbon in the biosphere. Wind energy is a result of thermal heating of the earth by the sun, having global patterns of a semi continuous nature. Geothermal renewable energy originates from heat stored beneath the surface of the earth. The source of this energy is from the earth's molten interior and the decay of radioactive materials. Solar energy is a result of radiation from the sun. Another form of renewable energy which has great potential in future is industrial waste heat. This form of energy is a result of unused heat streams from industrial processes. Manufacturing and processing industries such as Paper and Textiles are one of the major sources of this kind of renewable energy.

By definition, renewable energy sources should provide a continuous and unlimited supply of energy in particular to rural areas. However, several barriers are hindering promotion and penetration of its use. Barriers such as technical difficulties, the intermittent nature and some of the renewable energy sources, as well as constraints still pose limits to their wide promotion and deployment.

It is the fact that renewable energy sources are almost an unlimited supply of energy if one considers the energy required by mankind compared with the extremely large amount of energy we receive from the sun. For sustainable development, modern energy services require the growing inclusion of renewable energy into the sustainable energy mix of the country.

The technologies used now and in the future for conversion of renewable energy sources to heat; electricity and or fuels are plentiful in the country. These technologies can play part and contribute to the energy mix of the country. Their development will contribute to the gradual lowering of technology prices on the one hand and to improvement in their efficiency on the other hand. In the future, it is anticipated that renewable energy and its different energy conversion technologies will become economically viable, capable of competing with fossil-fuelled technologies in the Tanzanian market. However, this will succeed only if all the barriers will be tackled.

In the country there are several leading renewable energy sources which can be used in generating electricity in particular to rural areas. In the following sub-sections, the leading sources are discussed in detail. The information from these sub-sections was obtained from the assessment conducted from 2006 to 2010[Kusekwa et al., 2007].

4.1 Biomass energy

Energy consumption in the Tanzanian households accounts for more than 88 percent of the total energy, most being biomass. The trend is not expected to fall in the near future but to continue increasing as demand of energy increases. The increase is attributed to low pace of rural electrification caused by high cost of connection material, labour and high cost connection fees charged by the utility company which the majority of the rural poor population cannot afford. In this aspect, only biomass is still serving as the only affordable source of energy. However, utilization of conventional biomass is still high in most rural areas i.e. direct use of firewood, dung or semi processed in the form of charcoal. In this way, there is a need of sensitization to the population to use the available technologies or develop modern technologies which will be of great beneficial to the user.

Thus, new technologies or improving the existing ones have to be undertaken to add value to raw biomass and discourage the user to continue using the conventional methods.

Biomass sources suitable for energy generation in Tanzania covers a wide range of materials from firewood collected in farmlands; natural woods from agricultural and forestry crops grown specifically for energy generation or other purposes; crop residues and cow dung. It includes solid waste, timber processing residues etc. The most significant energy end-user is cooking and heating. During the assessment process, it was established that biomass sources can be divided into four major categories:

- Wood, logging and agricultural residue
- Animal dung
- Solid industrial waste
- Landfill biogas

It was noted that landfill biogas generation is dependent on environmental consideration and waste management practices in particular in the semi-urban and urban areas. The potential for exploitation of this source of renewable energy is high and will continue increasing in the near future because more and more people are migrating to semi-urban or urban areas where they consider opportunities for getting jobs and having good life. The semi-urban areas are now changing into big towns and the cities are growing and becoming bigger and complex. Hence, more wastes are expected to be generated daily.

Biomass is one of the renewable sources capable of making contribution to the future Tanzanian energy supply as well as contributing in job creation and hence poverty alleviation. During assessment process it was established that there are several forms in which biomass can be used for energy generation. Three sources are common i.e. residue, natural resources and energy crops. Residues are divided into three categories. The categories analysed are given in Table 6.

4.1.1 Natural sources
Natural sources include biomass gathered from natural resources such as fallen tree branches, woody weeds, etc.

4.1.2 Energy crops
Energy crops include biofuel as sole or principal product such as trees, grasses, and sugarcane, sorghum and oil crops. In addition, biofuel co-production is also part of energy crop category. Biofuel-co-production is a pre-planned multi-output production including biofuel i.e. sugarcane to produce sugar, ethanol, electricity, timber or tree-fruit production to deliver thinning and harvest waste as biofuel.

Generation of biofuel is expected to increase in the near future. A policy for biofuel has been developed by the government. The government is keen on development and generation of biofuel for the benefit of the country. More local or international investors are expected to participate fully in the production of biofuel and thus enhance the energy mix of the country. Availability of renewable energy sources varies depending on their attractiveness to the end user. Biomass differs markedly from conventional fuels and other renewable sources by having a wide range of competing use such as food, fodder, fibre, agricultural fertilizers, fuels, etc. In many places, some types of biomass are less valuable as resource energy than as source fulfilling other needs.

Primary residues	Secondary residues	Tertiary residues
Primary residues materials are usually from forestry, agricultural crops and animal rising. Primary residues can be categorized either as residues arising in concentrated form(dung from stalled livestock, harvested cereal straw, stalk, husk) or residues that must be gathered together (dung from grazing livestock, crop residues which are not harvested such as cotton and maize stalks)	Include material from: • Processing wood • Food and organic materials in concentrated form suck as • Sawmill bark • Tree chips • sawdust	Include waste arising after consumption of biomass such as sewage, municipal/city solid waste, landfill gas etc.

Table 6. Types of Biomass Supply

Potential of biomass sources (non-wood) in the country are given in Tables 7.

S/No	Renewable Energy Sources	Estimated Potential [MW]	Remarks
1	Sawdust	100	More studies are required to establish actual value
2	Sisal Residue	500	Will increase in near future
3	Crop residue	212	Initial estimation. Expected to increase
4	Cattle, Pig dung	-	More studies are required to establish actual value
5	Bagasse	57	Initial estimation. Expected to increase.
	TOTAL	**869**	

Table 7. Non-Wood Biomass Resource

Estimated average annual production levels of wood fuel and its associates such as tannin residue are shown in Table 8.

S/N O.	Renewable Energy Sources	Estimated Potential [Mw]	Remarks
1	Forest residue	523	Initial estimation. Its value could be high.
2	Wattle residue	15	Initial estimation
	TOTAL	538	

Table 8. Wood Biomass Resource

4.2 Solar energy

Solar radiation is the type of energy which is available at any location on earth. Solar energy in the country was assessed using the following criteria:

- power density or irradiance
- angular distribution
- spectral distribution

The maximum power density of sunlight on earth is approximately 1 kW/m² irrespective of location of the area or country. Solar radiation per unit area during a period of time can be defined as energy density or insolation [Renewable Energy Project Handbook, 2004]. Solar radiation is measured in a horizontal plane; the annual insolation varies by a factor of 3 from roughly 800 kW/m²/ year in northern Scandinavia to a maximum of 2,500 kW/m²/year in some desert areas such as: Kalahari etc. Practical applications of solar energy the absolute value yearly insolation is less important than the difference in average monthly insolation values. However, the differences vary greatly from about 25 % close to the equator, to a factor of 10 [Renewable Energy Project Handbook, 2004] in the most northern and southern areas. The average power density of solar radiation is normally 100-300 W/m² and the net plant conversion efficiencies are typically 10 % or less, hence, substantial areas are able to capture and convert significant amount of solar energy for energy generation. Tanzania is well situated near the equator; the country can capture and utilize solar energy in the purpose of rural electrification.

Solar energy presents great development in the country. Investigation conducted by Nzali et al [Nzali et al., 2001] suggested several areas in the country which can contribute to development of solar energy. Table 9 gives the insolation levels values in some areas of the country captured by the study. Solar photovoltaic energy is uniquely useful in rural not served by the National grid to provide basic services such as irrigation, refrigeration, communication and lighting. Solar energy is often more efficient than traditional sources such as kerosene. For lighting, a photovoltaic compact fluorescent light system is more efficient than kerosene lamp; used in rural areas to provide night lighting. Photovoltaic system also avoids the high costs and pollution problem of standard fossil-fuel power plant.

4.3 Wind energy

Wind is widely distributed energy source. Between 30°N and 30°S, air is heated at the equator rises and is replaced by cooler air coming from the South and the North. At the earth's surface, this means that cool winds blow towards the equator. Tanzania is situated near the equator; it is affected with the movement of the air movement as well as benefits from this prevailing condition.

The availability of wind varies for different regions and locations. It should be noted that mean wind speed may differ by as much as 25% from year to year. In some areas there are also significant seasonal differences. It has noted that in the country, there is a period when wind speeds are higher and some period wind speeds are low. Due to seasonal variations, the potential of wind for power generation can be significantly higher than the annual mean wind speed would indicate. Thus, not only the mean wind speed but also the wind speed frequency distribution, commonly described by a Weilbul distribution have to be taken into account in order to estimate accurately the amount of electricity to be generated. Wind speed varies with height, depending on surface rough ness and atmospheric conditions. Daily and hourly variations in the wind speed are also important for scheduling the operation of conventional power plant and adjusting their output to meet these variations.

Station	MONTHS											
	Jan	Feb	Marc	April	May	June	July	August	Sep.	October	Nov.	Dec
Dodoma	6.1	6.0	6.1	5.7	5.6	5.8	5.7	6.0	6.3	6.4	6.5	6.2
D'Salaam	5.2	5.3	4.9	4.0	4.3	4.4	4.4	4.0	4.9	5.1	5.8	5.6
Iringa	6.0	6.1	5.7	5.9	6.2	6.3	6.1	6.6	6.7	7.0	6.7	6.2
Kigoma	4.3	4.5	4.9	4.3	4.4	4.8	4.3	4.9	4.9	4.7	4.1	4.3
Mtwara	4.4	4.6	4.3	4.0	4.4	4.4	4.5	4.6	4.9	4.9	5.2	4.8
Musoma	5.4	5.0	5.4	5.4	5.4	5.0	5.2	5.4	5.4	5.4	5.7	5.4
Same	5.6	5.5	5.6	4.7	3.6	3.8	4.0	4.1	4.6	5.0	5.4	5.6
Songea	4.2	4.3	4.2	3.9	3.9	3.6	3.7	3.9	4.4	4.5	4.5	4.4
Tabora	5.6	5.5	5.8	5.4	5.6	5.5	5.1	5.7	5.6	6.0	5.2	5.4
Zanzibar	5.1	5.2	4.9	4.2	4.4	4.7	4.5	4.8	5.1	5.3	5.0	5.0

Table 9. Mean monthly Daily Insolation totals in kWhm^2/day for period of ten years [source A.H. Nzali 2001]

Wind resources can be exploited mainly in areas where wind power density is at least 400 W/m^{2}at 30 metres above the ground. Continuing technical advances has opened up new areas to development, Because of the sensitivity of the potential of the value of the wind speed, the determination of specific sites for wind energy projects depends on accurate meteorological measurements, and sites measurements etc. Even in the best sites, the wind does not blow continuously. Thus, it can never achieve the 100% required for electricity generation. Wind energy potential in Tanzania, wind power densities are given in Table 10.

Wind farms for commercial plants appear promising at Makambako and Kititimo in Singida region as well as Mkumbara, Karatu and Mgagao. Areas along rift valleys, the southern high lands and along Lake Victoria are reported to have some possibilities of potential wind sites.

Over the years, wind energy resources in the country have been used for wind mill to pump water. Less was been done in electricity generation. However with the availability of policy and renewable energy promotion program, emphasize now is toward utilization of wind energy in electricity generation. Number of wind mills available in the country is given in Table 11 and a photo depicting a wind turbine in Itungi village in central Tanzania is shown in Figure 1. The wind turbine is used to generate electricity for water pump.

Fig. 1. Wind Turbine used to Generate Electricity for Water Pumping

4.4 Geothermal energy

Geothermal energy tends to be relatively diffuse in nature that is why it is difficult to tap. Geothermal heat is concentrated in regions associated with the boundaries of tectonic plates in the earth's crust. Eastern lift valley and Western part of lift valley is the area where availability of geothermal sources has been located. It has been established that on average, the temperature of the earth increases by about 3°C for every 100m in depths.

The potential of geothermal is highly dependent on the results of the resources exploration survey, consisting the location and confirmation of geothermal reservoir, with economically exploitable temperature, volume and accessibility. There is some potential of geothermal resource in the country. Currently, the existing potential is being assessed by the government through the Ministry of Energy and Minerals (MEM). A geological survey to establish the potential has been conducted since 2006. The project is assessing the geothermal potential at Songwe west of Mbeya city, Southern Highland. The estimated geothermal potential is about 1,000 MW. Geothermal power is relatively pollution free energy resource which can contribute much to the energy mix of the country if commercially exploited

Wind Power Class	Wind Power Density, [W/m]	Wind Speed [m/s]	Wind Power Density W/s	Wind Speed [m/s]	Wind Power Density, [W/m]	Wind Speed [m/s]
1	100	4.4	160	5.1	200	5.6
2	150	5.1	240	5.9	300	6.4
3	200	5.6	320	5.5	400	7.0
4	250	6.0	400	7.0	500	7.5
5	300	6.4	480	7.4	600	8.0
6	400	7.0	640	8.2	800	8.8
7	1000	9.4	1600	11.0	2000	11.9

Table 10. Wind Power Densities [Source Mmasi et al., 2001]

4.5 Industrial Heat Recovery Power (IHRP)

Industrial heat recovery power represents a poorly known as renewable energy resource in the country, often unused and hence, often wasted resource in energy intensive industries. This resource can provide fuel-free electricity but has been neglected.

Industry heat recovery power use a wide variety of heat resources in applications such as cement, waste incinerators, pulp and paper mills, oil refineries, etc. The industrial applications for waste heat recovery do not require new sitting; the power unit can be installed within the boundaries of existing industrial site. IHRP does not influence the industrial process and does not interfere with the basic objective of production.

IHRP is not well known in the country, however, with the existing three cement industries, one paper mill (Mufindi Paper Mill) and Tipper oil refinery if harnessed they can contribute to the energy mix available in the country.

Region	Number of Wind Mills
Singida	36
Dodoma	25
Iringa	16
Shinyanga	6
Tabora	4
Arusha	4
Kilimanjaro	1
Mara	8

Table 11. Number of Wind mills in Tanzania (Source: Renewable Energy in East Africa – 2009)

4.6 Mini-hydropower sources

Out of estimated 315 MW small hydro potential in Tanzania less than 8 MW have been exploited by installing two power plants. The Ministry of Energy and Minerals (MEM) through REA has been funding studies for small hydro power plants. Dar es Salaam Institute of Technology (DIT) has participated in conducting these studies covering several villages, district, and regions with potential of small hydro power plant development. The villages, district and regions visited include Ruvuma, Rukwa, Iringa, Kagera, Morogoro,

Mbeya, Kigoma and Njoluma.   Identified potential river sites for small hydro power generation are given in Table 12. Assessments of actual power available from the established sites are still being worked out. However, the established sites have the potential of generating enough electricity to spur rural electrification in the identified areas. Water falls from the identified area is shown in Figures 2 and 3.

Fig. 2. Water fall at Madaba in South-Western Tanzania

Fig. 3. Water falls for mini-hydro power at Chita-Kilombero

Renewable energy exploitation in the country is still at an initial stage with a limited number of project developers, promoter's finance providers; services contribute less than 1% of the energy balance. Biomass within the renewable energy section accounts for more than 89% of the cooking resource in rural Tanzania, but the budget allocated by the Government for renewable energy services including biomass is limited to less than 1% of the annual energy development budget of the Ministry of energy and Minerals (MEM).

Nevertheless, renewable energy applications in the country have a good potential for powering development goals considering their local availability potential, the limited energy per capital consumption and ever-hiking prices of imported fossil-fuel.

Renewable energy will be a catalyst of rural development in the near future. It will play a major role in generation of electricity to spur quick rural electrication. However this, will be accomplished if the existing technologies are improved and new affordable technologies are developed. The following technologies are result of the assessment process conducted in the country from 2006 to 2010 by the author. Some of the technologies are old but need improvement to increase their efficiencies. New technologies need testing and commissioning.

S/No	Site	River	Load Centre	Head[m]	Discharge [m³/sec]	Capacity [kW]
1	Sunda Falls	Ruvuma	Tunduru	13.5	26	2x3,000
2	Kiboigizi	Kitanga	Karagwe	90	3.8	3,200
3	Kenge	Ngono	Bukoba	10	24	2,400
4	Luamfi	Luamfi	Namanyere	40	9	1,200
5	Mkuti	Mkumti	Kigoma Rural	23	3.3	650
6	Nakatuta	Ruvuma	Songea	67.8	50.3	1,500
7	Mtambo	Mtambo	Mpanda	17	13.5	2,000
8	Lumeme	Lumeme	Mbinga	301.2	1.31	4,200
9	Ngongi	Ngongi	Ruvuma	270.7	1.09	3,100
10	Luwika	Luwika	Mbamba bay	359.5	1.5	5,800
11	Mngaka	Mngaka	Paradiso	15	7.64	900
12	Songwe	Songwe	Idunda	75	1.5	720
13	Mngaka	Mngaka	lipumba	25	4.424	870
14	Kiwira	Kiwira	Ibililo	20	10	1,350
15	Prison	kiwira	Natural Bridge	30	12	3,000
16	Kitewaka	Kitewaka	Ludewa Township	50	9.884	4,200
17	litumba	Ruhuhu	Litumbaku Hamba	8	59	4,000
18	Mtigalala Falla	Lukose	Kitonga	70	10	5,000
19	Kawa	Kawa	Kasanga/Ngorotwa	65	0.3	130
20	Ijangala	Ijangala	Tandala	80	6	500

Table 12. Identified Potential River sites [Source REA-March 2010]

5. Renewable energy technologies (existing and new) in Tanzania

Renewable energy technologies deployment in the country is at initial stages of development, although it is not well quantified and well documented. The energy policy focuses on renewable deployment on biomass, solar, micro, mini and small hydro power plants and wind since it was felt that technologies for this energy sources could be disseminated in short term. Geothermal, with existing potential of about 1,000 MW exploitation is considered a long term option since the cost of its development is comparably high.

The use of energy sources such as solar, biogas and LPG especially in the household sector is still low. However, awareness is growing and it is anticipated that in near future its use will increase. It is estimated that about 1.2 MWp of photovoltaic power has been installed in the past three years for various power applications of which more than 35 percent of total installed capacity is from solar home systems (SHSs). The average sales of equipment relating to SHSs between 2000 and 2005 were about 500-600 PV systems per annum. The trend of sales in recent years is growing fast.

5.1 Biogas technology

Recent studies show that, more than 6,000 domestic biogas plants have been built countrywide for domestic and commercial applications. However, as these new technologies get rolled out to more remote areas, especially biogas they invariably encounter more isolated local cultures. For example in predominantly Muslim households it is difficult to convince the community to use pig dung to generate energy. Studies have revealed that pig dung is more efficient fuel than cow dung.

Hundred of tones of livestock dung across the country generated by cattle went unused every year, adding to that for example, The National Ranching Company Ltd (NARCO), has 10 ranches with about 33,000 animal units and proximity to around 55 Villages with a total population of around 156,900 individuals, who are also engaged in the livestock industry. With new innovations in more effective way bio-mass and bio-fuels, the hundreds of tones of cow-dung left over on the grazing land is a resource which could make a difference in the livelihood of the communities close to the ranches as a source of energy and fuel.

The use of bio-gas will reduce deforestation which contributes to global warming, leads to reduction in rains thus leading to low crops and vegetation growth and eventually reduction in crop and livestock production.

Biogas is a cheap [source of energy] when compared to other sources because it uses organic matter such as vegetables and animal waste. Bio gas turned into electricity will improve the quality of life for communities within and around the ranches. Biogas project helps to reduce waste, bacteria and waste odour and clean up the environment. Bio gas based electricity could be linked with solar powered electricity as a hybrid system in order to promote decentralized power systems and consequently enhance energy security.

Dar es Salaam Institute of Technology (DIT) has developed a portable biogas plant made from plastic containers which can be used by rural households. The scheme is shown in Figure 4 and is cheap and affordable. Besides DIT, Small Industries Development

Organisation (SIDO), GAMARTEC, VETA, and private enterprises are researching and developing biogas plants for domestic and institution applications.

Fig. 4. A biogas plant using plastic containers (Source DIT R&PGS-2011)

Biogas is a feasible option for the domestic energy needs of Tanzania's rural population and offers the following socio-economic and environmental advantages

- provides a low cost energy sources for cooking and lighting
- improves sanitation in the home, farmyard and surrounding environment
- eliminate respiratory and eye diseases caused by indoor air pollution
- save time for women and children because they don't need to collect firewood
- create rural employment
- reduces greenhouse gas emission
- reduce deforestation
- produces an effluent called bio-slurry which is an excellent organic fertilizer

5.1.1 Improved stove technology

Tanzania has about 35 million hectares of forests; of which about 38 percent of total land areas (13 million hectares) are protected forest reserves and the remaining 62 percent are forests on public land in village areas that are under pressure from human activities including harvesting for energy. Forest and trees in farmlands contribute to wood fuel supply. However, supply of wood fuel is declining rapidly in the country causing scarcity of energy to rural and semi-urban low-income families and environmental degradation in areas where harvesting of wood fuel exceeds the growing stock potential.

Much of the research and development work carried out on biomass technologies to serve the rural areas has been based on improvement of available traditional stoves. This was initially in response to the threat of deforestation but has been focused on the needs of women to reduce fuel collection time and improve the kitchen environment by smoke removal.

There have been many approaches to stoves improvement, some carried out by local institutions, individuals and others as part of wider programmes run by international organisations.

Some of the features considered in improving the stoves include:

- An enclosed fire to retain the heat
- Careful design of pot holder to maximise the heat transfer from the fire to pot
- Baffle to create turbulence and hence improve heat transfer
- Dampers to control and optimise the air flow
- A ceramic insert to minimise the rate of heat loss
- A grate to allow for variety of fuel to be used and ash to be removed
- Metal casing to give strength and durability
- Multi pot system to maximise heat use and allow several pots to be heated simultaneously

Designs of stoves depend on the form of biomass providing energy. Improving a stove design is a complex procedure which needs a broad understanding of many issues. Involving users in the design is essential for a thorough understanding of the user's needs and requirements of the stove. The stove is not merely an appliance of heating food, but in rural context is often acts as a social focus; a means of lighting and space heating. Tar from the fire can help to protect a thatched roof, and the smoke can keep out insects and other pets. Hence, cooking habits need to be considered as well as the lifestyle of the users.

Fuels with improved designs of stoves include firewood, charcoal and sawdust. Based on the assessment conducted, it has been established that there are difference between stoves used in rural, urban and institutions or commercial ventures. Use of firewood is predominant in the rural areas and as one travels into the urban areas there is a shift to charcoal.

It was established during the assessment that stoves in use in rural areas are normally adaptable to using more than one form of biomass such as wood and agricultural wastes. Firewood is used widely in the rural areas. The traditional firewood stoves used in rural areas is normally at no cost to the user and these stoves have a lot of inefficiencies. One stove fits any size of pot and the intensity of fire is controlled by adding or removing fuel from the stove. The fuel i.e. firewood is not bought but collected free of charge from the forest or farms. Urban stoves are normally single fuel devices. Charcoal is a very important fuel for urban areas and is usually purchased rather than collected.

Improved stoves designs in the country to date are usually targeted to urban dwellers. This has been probably been due to the higher income levels of this group of people. Hence, improved charcoal stoves are widely disseminated stoves technology. Improved charcoal stoves are highly efficient stoves that save fuel and money because the heat to be lost is minimized by some insulation included in the design. These stoves can save about 35%-40% charcoal over traditional stoves.

Sawdust stove designs are also finding their way into the market especially in small business enterprises called "Nyama Choma" or meat roasting in the urban areas. There are some few problems that would need to be addressed in order to make the technology popular in the country.

Improved stove technology focuses on improving firewood consumption. In the long run it aims at reducing carbon dioxide emission and indoor air pollution, reducing workload to women and children and conserving forest resources. The overall aim of the project is to improve thermal performance of the woodfuel stoves in rural areas

Other benefits are income generation opportunities especially to village technicians. Stove improvement technology adds value on indigenous technology that uses indigenous fuel resources and material. Improved stove technology project is designed to start with small models that can be replicated in the whole country. The project will relate to construction of the efficient stoves and imparting knowledge on proper management of woodfuels. Amongst of the improved stove is *"Jiko Mbono"* shown in Figure 5. The stove is a Top-Lit-UpDraft (TLUP) gasification stove with natural draft air supply. The stove can use Jatropha seeds directly instead of Jatropha oil.

Fig. 5. Jiko Mbono Underdevelopment (Source DIT R&PGS-2011)

Timber and manufacturing industries in the country generate a lot of sawdust (shown in Figure 6). The sawdust can be used as a renewable source of technology. Sawdust stoves have been developed as can be seen in Figure 7. The stoves are cheap and affordable and can be used in both semi-urban and urban areas. The sawdust stoves are expected to be popular in future.

Fig. 6. Sawdust accumulation at one of Timber Processing Industry in Tanzania

5.2 Solar Technology

Over several past decades, new commercial industries have been established for an assortment of solar energy technologies, demonstrating schemes with wide variations of success. The SHSs system components are usually imported through various private sector initiatives. The common PV applications in the country are household lighting, telecommunication, vaccine refrigeration in rural and semi-urban areas, powering electronic accessories e.g. radios, TVs, computers etc, etc. water pumping, powering schools and health centres and rural dispensaries.

Dar es Salaam Institute of Technology (DIT) has developed a high power solar thermal system based on parabolic concentrator Heliostat. The scheme is cheap to construct and can be used by institutions in the country. The system is capable of concentrating 20 kW per unit Heliostat. The unit can be cascaded to a very high power station. A parabolic concentrator is given in Figure 8

5.3 Wind Technology

Based on Mmasi, Lujara and Mfinanga [Mmasi et al, 2001] on wind energy potential in Tanzania, wind resources are expressed in wind power classes ranging from class 1 to class 7, with each class representing a range of mean wind power density or equivalent speed at specified height above the ground. In this aspect, Mtwara, Dar es Salaam, Pwani, Tanga, Kigoma, Kagera, Singida, Dodoma, Tabora, Shinyanga, Morogoro, and part of Southern Arusha are suitable areas for future generation of electricity using wind as the source of energy.

Fig. 7. Sawdust stove in Rural Tanzania

Fig. 8. A Parabolic Concentrator Heliostat (Source DIT-R&PGS-2011)

5.4 Domestic waste technology

Household wastes can also be used as source input in generating a closed system steam known as electrical generator from a local kitchen. Small scale electrical generator capable of utilising wood stove waste heat has been developed by DIT. It is anticipated that the system will find market in the future. Figure 9 shows the 3D concept of the steam-electrical generator.

5.5 Cogeneration technology

There is few biomass based-co-generation plants in the country. These include sugar processing plants: Tanganyika Planting Company (TPC), Kilombero Sugar Company (KSC), Mtibwa Sugar Estates (MSE) and Kagera Sugar Company, Tanwat (Tanning) and Sao Hill Sawmill have waste that can be used in generating electricity. Table 13 shows electricity generated from cogeneration technology.

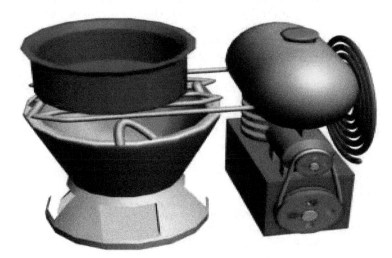

Fig. 9. The 3D Concept of the Steam –Electrical Generator Prototype (Source DIT R&PGS-2011)

The highlighted technologies have the potential to contribute to rural energy electrification in the country in the near future. Some of the technologies are still under development. Wind and solar technologies are already in application. These technologies need improvement. Tertiary, higher learning institutions and research centres in the county have the role of improving these technologies.

S/No	Plant Name	Main Resource	Region	Capacity [MW]
1	TANWAT	Biomass	Njoluma	2.7
2	TPC	Biomass (Bagasse)	Kilimanjaro	20
3	Sao Hill	Biomass(Saw Dust)	Iringa	16
4	Ngomeni	Biomass (Sisal Waste)	Tanga	0.5
5	Mwenga	Biomass (Sisal waste)	Tanga	3.36

Table 13. Electricity Generated from Cogeneration Technology (Source: MEM 2011)

With all of the technologies already in place and some underdevelopment, there are still many challenges and barriers that need attention. Some of the barriers are presented in section 6.

6. Barriers to promoting renewable energy technologies in Tanzania

Renewable energy technologies are still perceived as "niche" energy resources by many Tanzanian. Barriers to their enhancement and development are on all levels i.e. cognitive, perceptual, policy attitudes and in the economic sphere. Renewable energy technologies are perceived by many Tanzanian as complementary energy not main, hence, still in the learning curve phase in their developments. They are viewed as relatively new, not sufficiently field proven, somehow expensive to purchase, to install and to maintain. They are often viewed as small, dispersed resources, of unstable output, and incapable of providing sustainable energy for the future. They lack base expertise, information on cost is imprecise and thus there are high impediments to possible capital investment.

The economic barriers are both real and perceived. The real economic barrier is influenced by unfair competition from fossil-fuels or conventional energy sources. Economically, renewable energy technologies project suffers from high up-front capital requirements, high interconnection costs, and lack of financing mechanisms from financial institutions e.g. commercial banks, etc. Financial institutions in Tanzania still perceive investment in development of renewable energy technologies as high economic risk; their entire economic structure is viewed as poor, with long amortization.

The general barriers for development of renewable energy technologies are summarized according to resources as follows:

6.1 Biomass
- Dispersed form of energy,
- Variety of technological solutions
- Competition from higher value applications
- Not sufficiently mature, therefore, risk to investors
- Difficult due to collection in some areas and transportation
- In case of Bioenergy, it is land-intensive
- Low load factors, hence it tends to increase energy system costs
- Minor influence on Tanzanian energy supply
- Not modern enough for mass utilization

6.2 Wind
- Uncompetitive technology in the short and medium run
- Lack of good wind conditions i.e. speed in many part of the country
- Lack of financial resources to finance research and development of wind turbine in the country
- Lack of human resources for servicing and maintenance after installation of the system

6.3 Geothermal
- Drilling technology difficulties
- High up-front investment
- Resource handling problems such as resource depletion, corrosion, etc

- Financing constraints due to high up-front costs
- Competition from fossil fuel power plant

6.4 Solar
- Low energy density in some areas in the country
- Resource available only during daytime
- Sensible to atmospheric and weather fluctuations
- Higher cost of Solar PV
- High capital cost
- Long payback periods
- Grid connection issues
- Storage issues
- High cost of storage solutions
- Hazardous materials in PV systems (battery) etc
- Lack of financial capability to subsidise solar energy projects

6.5 Industrial heat recovery power
1. Lack of awareness of this unused form of renewable energy in manufacturing and processing industries
2. Not included in government energy master plan
3. Unawareness of waste heat potential for power generation in the country
4. Perception as nuisance not convergent with the basic function of manufacturing and processing process of the mills or factory
5. Fear of damage caused to the production process
6. No environmental credits given for waste heat power generation from waste heat by the government
7. Financing constraints because of high up-front costs
8. Lack of interest in using the waste heat potential for generation of electricity
9. Preference for external solutions such as diesel generator applications, etc.

7. Recommended actions to remove barriers to promoting renewable energy technologies

- Development of effective public awareness and promotion programs that depend mainly on market surveys and studies and concentrate on media in particular television programmes and newspapers.
- Allowing systems and spare parts of the developed technologies to be available in shops.
- Establishment of maintenance centres.
- Demonstrating developed technologies can be presented in international trade fairs, engineering conferences, municipals and city councils, big factories etc.
- Establish incentive mechanisms innovators and developers.
- Encourage local manufacturing companies to manufacture the systems.
- Form federation, union or society which bring together representatives of users, companies, financial sources, policy makers and researchers in order to coordinate efforts in using the developed technologies.

- Establish credit mechanism to finance prospective technologies.
- Establish a programme or mechanism to solve the problem of already installed systems. The programme should include some mechanism for informing the user about the systems and their regular duties.
- Setting up coordinating committee for planning and implementing the action plan to acceptable technologies.
- Strengthening the cooperation between the concern ministry, authorities, institutions and organisations involving them in the national action on renewable energy technologies.
- Setting rules and legislation for quality assurance, standardisation and certification for all renewable energy technologies components and systems.
- Development of effective public awareness and promotion programmes such as demonstrating systems, some printed materials (leaflets, brochures etc) training courses, seminars, presentations and workshops for targeted users, small-scale laboratories in schools, technical colleges and universities.

8. Social impact of renewable energy technologies in sustainable development

Renewable energy resources and technologies can serve as one of the key drivers for rural development in the country in a number of ways in:

- Enhancing local micro-economic development in agriculture, manufacturing, and small industries
- Providing vital economic generating activities in the rural areas such as water pumping, battery charging, lighting schools, ICT development, crop drying, milk refrigeration, drug refrigeration, and ice making in semi-urban areas.
- Improving human development such as accessibility to modern education, internet, and improve health services, etc.
- Helping to lower the pace of migration of young people to overcrowded municipalities and cities.
- Preventing social unrest in particular to young people
- Poverty alleviation

Renewable energy technologies for rural and semi-urban electrification is more sustainable; suitable for supplying geographically dispersed villages by means of distributed energy often without relying on a national grid. Grid connection to remote and dispersed villages is expensive and technically difficult; therefore, local mini-grids developed from renewable energy sources can be established and serve the purpose of rural electrification either as stand lone power generating unit for a particular village or interconnected with other village generating unit. In this way:

- Biomass. The majority of the rural population in Tanzania relies on traditional biomass to meet their cooking and heating needs. The challenge is to ensure more efficient and sustainable use of biomass for heat extraction, cooking and generation of electricity instead of using raw biomass.
- Solar PV systems can be widely used in poverty alleviation projects for electrification of remote underdeveloped areas.

- Wind turbine can provide modern, clean, sustainable and economical energy for remote village areas, either via local mini-grids or as stand –alone option.
- Mini-hydro power plants can provide modern, clean and sustainable and economical energy.
- Geothermal energy can provide sustainable continuous energy, independent of weather conditions.

Poverty alleviation objective in the country focuses on rural areas. Hence, development of renewable energy technologies and infrastructure can effectively contribute to poverty alleviation in the coming years.

9. Distribution generation

Dondi *et al* (2002) defined distributed generation as a small source of electrical power generation or storage ranging from less than a kW to tens of MW that is not a part of large central power system and is located close to the consumer (load). Chambers (2001) also defined distributed generation as a relatively small generation units of 80 MW or less. According to Chambers, these units are sited at or near customer sites to meet specific customer needs, to support economic operation of the distribution grid or both. The two definitions assume that distributed generation units are connected to the distribution network. It is clear that the two definitions give or allow a wide range of possible generation schemes. So, the definitions allow the inclusion of larger scale generation units or large wind farms, landfills, etc connected to the transmission grid, others put the focus on small-scale generation units connected to the distribution grid. Nevertheless, all the definitions suggest that at least the small scale generation units connected to the distribution grid are to be considered as part of distribution generation. Moreover, generation units installed close to the customer (load) or at the customer side of the meter are also commonly identified as distributed generation.

Ackerman *et al* (2001) precisely defined distributed generation in terms of connection and location rather than in terms of generation capacity. They defined a distributed generation source as an electrical power generation source connected directly to the distribution network or on the customer side of the meter. The definition is adopted in this chapter even though it is rather broader. The definition does not put limit or technology or capacity of the potential distributed generation application. It suits the Tanzanian condition when referring to renewable resources the country has.

Distribution generation is a latest trend in the generation of electrical power. The distributed energy resource concept allows consumers who are generating electricity for their own needs to send surplus electrical power back into the power grid or share excess electricity via a distributed grid. Distributed generation system can be divided in two segments, as shown in Figure 10. The segments include:

- Combined heat and power
- Renewable energy resources

Combined heat and power (CHP) is the use of a power generator to simultaneously generate both heat and electricity. The method is new in the country, but can be applied at Mufindi Paper Mills (MPM) where the paper machines generate enough back pressure which is useful in electricity generation.

Renewable energy resources (RER) capture their existing flow of energy, from on-going natural processes such as solar, wind, small hydro and biological processes. The two segments are the main component in implementation of distributed generation in the country and can accelerate rural electrification, hence improving accessibility of Tanzanian to modern energy and spur sustainable development.

Distributed generation could serve as a substitute for investment in transmission and distribution capacity or as a bypass for transmission and distribution costs. Distributed generation could result in cost savings in transmission and distribution of about 30% of electricity cost compared to the cost incurred by TANESCO in electrifying rural areas. Distributed generation can substitute for investments in transmission and distribution capacity. It can be used as an alternative to connecting a customer to the grid in a stand alone application. Furthermore, well selected distributed generation from the resources the country has can contribute in reducing national grid losses.

Distributed generation can contribute in the provision of ancillary services. These include services necessary to maintain a sustained and stable operation of the national grid. For instance to stabilize a dropping frequency due to a sudden under capacity such as power plant switching off due to technical problems or excess demand.

Installing distributed generation schemes will allow the exploitation of cheap fuel resources available in the country. For example in the proximity of landfills resources, distributed generated units could burn landfill gases. Also, biomass resources may be envisaged.

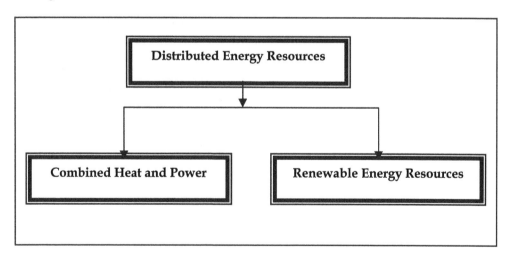

Fig. 10. Split up of Distributed Energy Sources

Increased environmental concern induces an increase interest in distributed generation application worldwide, also in innovations in the appropriate technologies. Nevertheless, the economic as well as technical challenge will be to optimally integrate the distributed generation units in the electricity system available in the country that up to now has been very centralized. The challenges emanating from application of distributed generation in the country is a blessing as more opportunities to engineers and researcher in investigating new and affordable means of integrating distribution generation in the national grid.

10. Discussion

A suite of off-grid renewable energy technologies such as solar home systems, biogas cook stove and either mini or small hydropower plants can deliver energy services to rural households most cost effective than national grid, without relying on expensive and polluting fuels. These Renewable Energy Systems are making inroads in some regions worldwide. But assessments on how they perform often neglect local realities.

In conducting the reviews, it has been revealed that cultural attitudes and social expectations can prevent the use of renewable energy technology applications as significantly.

11. Conclusion

The great challenge in Tanzania with regards to provision of modern energy i.e. electricity is found in the rural areas. Rural area development should be the overall priority in the energy access challenge, with the focus on increasing investment from government and private sector as well as assistance from development partners. Greater effort should be made to promote renewable energy technologies and energy resources available in the country. Tanzania is rich in renewable energy resources; however, due to lack of financial resources, wood fuel and human physical power will continue to be the main source of energy in most rural areas of the country. In order to bring sustainable rural development, it is crucial that long term efforts scaled up to facilitate access to modern energy services in rural areas by employing renewable energy technologies. Rural areas of Tanzania need to use more modern energy services if poverty is to be made history. More modern energy is required to increase and improve production of value added goods and services in rural areas and this can be generated by new and improved renewable energy technologies.

The generation and use of modern energy in the country is a problem of capital and technology, not energy resources. It is a question of whether or not adequate capital and technology would be availed to develop the available renewable energy resources. Hence, the government, development partners and private sector, need to urgently avail the necessary capital to develop and generate own energy from the available renewable energy resources. At the same time, engineers, technologist and private firms corporate in research and development of technologies suitable for the country. The approach will bring positive economic performance and sustainability.

12. Recommendation

It is recommended that for all abundant renewable resources existing in the country and the available renewable energy technologies together will play a major role in the provision of affordable modern energy to rural areas. However, the following issues should be addressed:

- Reduce the initial high cost of implementing renewable energy projects
- Promote relevant research and development in renewable energy technologies

- Disseminate the available renewable energy technologies to the public
- The use of Biogas digesters and sawdust stove should be popularized, by the government. Research centres in the country should be encouraged to design effective digesters as this will help utilize the energy stored in cow dung.
- Energy stored in wind can be tapped and used for pumping water and electricity generation using wind. The government must work in collaboration with experts in this field to built sufficient wind mill to alleviate the energy problem.
- Tanzania has enough coal deposit. The use of coal oven should be highly encouraged as this will solve the energy needs of the rural dwellers.
- Establish energy data bank that will be a model for addressing nation's energy needs.
- Develop local capacity on use of renewable energy technologies.
- Introduce and application of distributed generation, which could serve as a substitute for investments in transmission and distribution capacity.

13. Indexes

Calorific value conversion factors for wood fuel (firewood) at final user level

1 kg of firewood	=	13.8 MJ
1 kg of charcoal	=	30.8 MJ
1 m³ of solid wood	=	0.725 Tonne
1 m³ of wood	=	10,000GJ
1 kWh	=	3.6 MJ

1 Tonne of fibre of sisal produces 25 tonnes of residue
1 Tonne of sugar produces 5 tonnes of bagasse
Energy content of wood fuel (air dry, 20% moisture) = 15GJ/t
Energy content of Agricultural residue (range due to moisture content) = 10-17GJ/t
Energy cost: 1 GJ costs U$ 0.95

4 kg of Jatropha beans	=	1 litre of biodiesel
10 tonnes of Jatropha beans	=	2,500 litres of biodiesel

14. Abbreviation

DRC	Democratic Republic of Congo
TANESCO	Tanzania Electricity Supply Company Limited
MW	MegaWatt
EWURA	Energy, Water Utilities Regulatory Authority
NEP	National Energy Policy
RE	Renewable Energy
RT	Renewable Technologies
R&D	Research and Development
REA	Rural Energy Agency
REF	Renewable Energy Fund
PSMP	Power Sector Master Plan
kW	kilowatt
IHRP	Industrial Heat Recovery Power
DIT	Dar es Salaam Institute of Technology
R&PGS	Research and Postgraduate Studies

SHS	Solar Home Systems
NARCO	National Ranching Company Limited
SIDO	Small Industrial Development Organization
VETA	Vocational Education Training Authority
TLUP	Top-Lit Up Draft
TPC	Tanganyika Planting Company
KSCL	Kagera Sugar Company Limited
KSC	Kagera Sugar Company
MSE	Mtibwa Sugar Estate
TANWAT	Tanganyika Wattle Company
ICT	Information and Communication Technology
CHP	Combined Heat and Power
MPM	Mufindi Paper Mills
RER	Renewable Energy Resources
CAMARTEC	Centre for Agricultural Mechanisation and Rural Technology
IT	Information Technology
URT	United Republic of Tanzania
FAO	Food and Agriculture Organization
UNDP	United Nations Development Programme
SIDA	Swedish International Development Agency
TaTEDO	Tanzania Traditional Energy Development and Environment Organization
IPPs	Independent Power Producers
EC	European Commission
IEEE	Institution of Electrical & Electronics Engineers

15. References

Casmiri Damian (2001). *Energy Systems: Vulnerability-Adaptation –Resilience. Regional Focus: Sub-Saharan Africa-Tanzania.* 56, rue de Passy-75016 Paris-France

Magesa Finias (2007). *Country Chapter: Tanzania.* Deutsche Gesellschaff fur Technische, 65760 Eschborn Germany

URT (2007) *Report- Tanzania Investment Centre.*

URT (2009). *Report on Economic Survey*

Msaki, P.K. (2006). *The nuclear Energy Option for Tanzania: A development vision for 2035, in Energy Resources in Tanzania, Volume I,* Tanzania Commission for Science and Technology.

Nzali A.H. (2001) *Insolation Energy Data for Tanzania,* International Conference on Electrical Engineering and Technology, The University of Dar es Salaam. Pp.EP26-EP32

Mmasi, R.C., Lujara, N.K., & Mfinanga, J.S. (2001), *Wind Energy Potential in Tanzania,* International Conference on Electrical Engineering and Technology, The University of Dar es Salaam, Pp.EP6-EP11

Karekezi, S. and Ranja, T., *Renewable Energy Technologies in Africa,* AFREPEN, 1997

Kristofen L. A., and Bokalders V., *Renewable Energy Technologies-their Application in Developing Countries,* IT Publications, 1991

Westhoff, B. and Germann, D., *Stove Images,* Brades and Aspel Verlag GmbH, 1995

Stewart et al., *Other improved Wood, Waste and charcoal Burning Stoves,* IT Publications, 1987

Vivienne et al., *How to make an Upesi Stove: Guidelines for Small Business,* IT Kenya, 1995

Lyidia, M and Mary, S., *Appropriate Household Energy Technology Development Training Manual,* IT Kenya, 1999

Caroline, A. and Peter, Y., *Stoves for Sale: practical Hints for Commercial Dissemination of Improved Stoves,* IT, FAO, IDEA, GTZ, FWD, 1994

Daniel et al., *Smoke Health and Household Energy Volume 1: Participatory Methods for Design, Installation, Monitoring and Assessment of Smoke Alleviation Technologies,* ITDG, 2005

UNDP/World Bank, ESMAP Project, *Sawmill Residue Utilisation study,* 1988

Renewable Energy Directive 2001/77/EC

http://www.iea.org "Distributed Generation in Liberalised Electricity Market

http://www.electricitymarkets.info/sustelnet

Ackermann, T., Anderson, G., and Soder, L. (2001), *Distributed Generation: a definition,* Electric Power systems Research, Vol. 57, pp. 195-204

Chambers, A., (2001), *Distributed Generation: A Nontechnical Guide,* PennWell, Tulsa, Oklahoma, pp.23

Dondi, P., Bayoumi, D., Haederli, C., Julian, D., and Suter, M., (2002), *Network Integration of Distributed Power Generation,* Journal of Power Sources, Vol. 106, pp. 1-9

IEA, (2002), *Distributed Generation in Liberalised Electricity Markets,* Paris, pp. 128

Voorspools, K., and D'haeseller, W., (2002), *The valuation of small Cogeneration for residential heating,* International Journal of Energy Research, Vol. 26, pp. 1175-1190

Voorspools, K., and D'haeseller, W., (2003), *The impact of the Implementation of Cogeneration in a given context,* IEEE Transactions on Energy Conversion, Vol. 18, pp. 135-141

DECON & SWECO (2005), Tanzania Rural Electrification Study, stakeholder Seminar no. 2, Impala Hotel- Arusha Tanzania, 20th April 2005

Kusekwa, M.A., Mgaya, E.V. and Riwa, A.A., (2004) *Micro-hydropower and Rural Electrification in Tanzania,* Yemeni Journal of Science Vol. 5, no. 2, pp. 91-108

MEM (2003), *The National Energy Policy, Ministry of Energy and Minerals,* The United Republic of Tanzania

Sawe, E. N., (2005), *Rural Energy and Stove Development in Tanzania, Experience, Barriers and Strategies,* TaTEDO

Sawe, E.N., (2005), *A Paper on Tanzania Energy Potential and Development Status/Energy consumption Summary*

Hifab International and TaTEDO (1998), *Tanzania Rural Energy Study,* TaTEDO, SIDA, MEM, Dar es Salaam Tanzania

Wamukonya, N., (2001), Renewable Energy Technologies in Africa: An overview of Challenges and Opportunities. Proceedings of the African High Level regional Meeting on Energy and Sustainable development, UNEP Collaborating Centre on Energy and environment

World Bank (2003), Little data book, Quick Reference to the World Development Indicator

3

Recent Developments in Renewable Energy Policies of Turkey

Hasan Saygın[1] and Füsun Çetin[2]
[1]Istanbul Aydın University, Engineering and Architecture Faculty
[2]Istanbul Technical University, Energy Institute
Turkey

1. Introduction

Nowadays, a radical change is taking place in global energy policies. A new energy paradigm consistent with the goal of sustainable development is evolving. The World is in the midst of paradigm shift towards non-carbon based economy. In nature of things, the new energy paradigm has emerged in and is taking root in developed countries. It is however spreading from them to developing countries. Renewable energy constitutes one of the three essential pillars of the new energy paradigm, due to its potentially important role in improving energy security and the decarbonization of global economy. One of the most important implications of this paradigm change is that technological leapfrogging opportunity appears for developing countries having sufficient renewable energy potential. Implementation of the new energy paradigm in developing countries can provide them to develop by avoiding from a repetition of the mistakes of the industrialized countries (Saygın & Çetin, 2010).

Turkey with huge renewable energy potential is one of these countries having a strong chance of leapfrogging in energy technologies. Although concerns about energy supply security dominates because of rapidly rising energy demand, decision-makers in Turkey are striving to set up secure, environment-friendly and sustainable energy policies parallel to contemporary global energy policies. In this context, it has been made important progress with regard to especially, renewable energy and energy efficiency regulations in the recent years (Saygın & Çetin, 2010). Present status and potential of renewable energy of Turkey and recent developments in its renewable energy policies are reviewed in the following sections.

2. Turkey's energy challenges and renewable energy

Turkey is 17th largest economy of the World. Although its energy use is comparatively low, the Country with rapidly growing economy is one of the fastest growing energy markets in the World. Primary Energy Demand is rapidly increasing, as can be seen from Figure 1. The Country will likely see the fastest medium-to-long term growth in energy demand among the IEA member countries. (IEA, 2009).

Turkey's total final energy consumption of energy was 74 Mtoe in 2008 up by %86 from 1990. Following its economic growth, energy use in Turkey is expected to roughly double over the next decade, and electricity demand is likely to increase even faster. This implies the needs for large energy investments but also measures for ensuring energy security, especially in electricity sector (IEA, 2009).

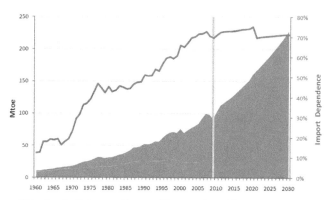

Fig. 1. Evolution of Turkey's Primary Energy Demand and Import Dependence (OME,2008).

Although Turkey is poor in hydrocarbons, its primary energy consumption is mainly based on fossil fuels as seen from Figure 2. Except hydro, renewable resources have been almost untouched up to recently. Under this circumstance, rapidly increasing energy consumption implies rapidly increasing import dependence, as seen also from Figure 1 including for practically all oil and natural gas and most coal. More than about 70% of the total primary energy consumption in the country is met by imports. It is heavily dependent on foreign fossil fuels and this dependency is one of the most important issue threatening its energy supply security and economy.

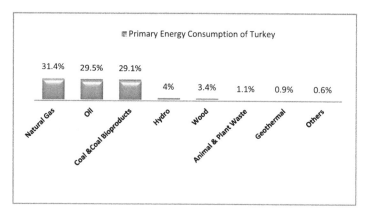

Fig. 2. Primary Energy Consumption of Turkey by Sources (MENR).

Concerns for ensuring sufficient energy supply for growing economy, therefore, dominates in determining energy policies. Hence energy security has a more central place in the energy policy goals in comparison with market reform and environmental protection. This fact may retard the diffusion of the new paradigm in the Country. Likewise, Turkey follows a deliberate policy for new renewables.

Another challenge rising from largely reliance on fossil fuels is rapidly increasing greenhouse gases emission. Although Turkey have less greenhouse gas emission per Capita than both OECD Countries and transition countries, it has a high rate of increase in emissions since 1990. Energy-related CO_2 emission has rapidly increased over last decade also, as illustrated in Figure 3. It is likely to continue to increase fast over the medium and long term, in parallel with significant growth in energy demand (IEA, 2009). This is a another growing concern in the Country.

Turkey has been a Party to the United Nations Framework Convention on climate Change (UNFCCC) since 2004 and to the Kyoto Protocol since 2009. Signing the Kyoto Protocol does not put an additional burden on Turkey until 2012. However, Turkey has undertaken the responsibility of passing the necessary legislation to lay the infrastructure for fighting climate change after 2012. The major issue for the Country is how to contribute to reducing the on global emission without jeopardizing its economic and social development prospects.

Its high energy intensity is another challenge for Turkey. The change in the primary energy density throughout the periods from 1980 to 2005 and from 2000 to 2008 are illustrated in Figure 4(a) and 4(b). In spite of improvement efforts, energy intensity remains high although an improving trend is observed currently.

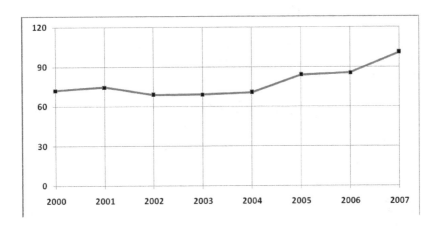

Fig. 3. The CO_2 Emission from Electricity Production (2000-2007) (MENR,2010).

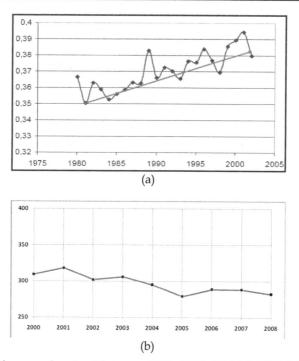

(a)

(b)

Fig. 4. Evolution of energy density a) between 1975- 2005 (TOE/ $ 1000(Çalıkoğlu,2007)
b) between 2000-2008 (kg equivalent oil/$ 1,000) (MENR,2010)..

As seen from above graphs, it has become near stagnant, after a few decades of rapid
increase. High energy intensity is a major obstacle in reducing emissions. But still, Turkey's
energy intensity is significantly higher as compared to the other OECD and IEA Countries
As seen from Figure 5 (Çalıkoğlu, 2007). That is, Turkey cannot use its energy efficiently.

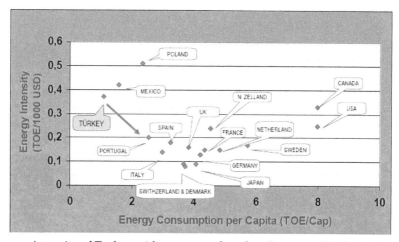

Fig. 5. Energy intensity of Turkey with respect to the other Countries (Çalıkoğlu,2007)

Total share of renewable in TPES has declined depending on, mainly, decreasing biomass use and the growing role of natural gas in the system. It was estimated that the share of renewable energy will decrease to % 9 of TPES in 2020 (IEA, 2005).

As seen from Figure 6, the share of installed renewable capacity in total installed capacity dramatically decreased in the last decade.

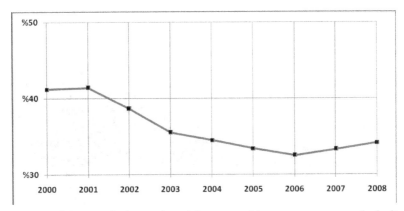

Fig. 6. The Share of the installed capacity of the renewable energy resources including large hydro, within the total installed capacity (%) (MENR, 2010}.

Although the absolute value of renewable energy use grows, since it doesn't grow at the same proportion with energy consumption, the share of fossil fuels continues to increase (Saygın & Çetin, 2010). This fact can also be seen from Figures 7-8 illustrating Turkey's primary energy supply and electricity generation by fuels.

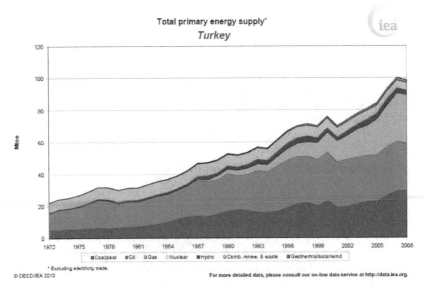

Fig. 7. Turkey's total energy supply by fuel in the period 1972-2008 (IEA, 2009).

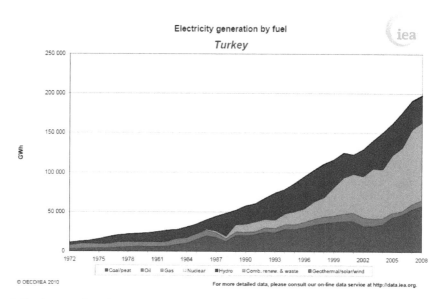

Fig. 8. Evolution of Turkey's electricity generation by fuels (IEA, 2009).

Environmental and energy security risks are therefore more and more increasing for the Country. The energy situation of Turkey is characterized by high rate of import dependence, high energy intensity, rapidly rising greenhouse gas emissions. This situation is evidently unsustainable and in conflict with the contemporary global trends. Turkey is faced with serious environmental and energy security challenges presently. Additionally, it needs for high amounts of financial resources for energy investment to meet growing energy demand. Limitations within the scope of harmonization with the EU cause the cost of energy investments to rise and complicate the situation further. All of these are matters of concern for decision makers of the Country (Saygın & Çetin, 2010).

It is clear that, the existing renewable energy potential should be realized in a reasonable time period. To realize renewable potential, Turkey has taken some steps in the right direction. Although it has been made important progress with regard to renewable energy, energy efficiency and market liberalization regulations in the recent years; new paradigm is slowly diffusing into Turkey's energy policies and the regulations due to strong concerns related with energy supply security[(Saygın & Çetin, 2010).

3. Recent developments in renewable energy policy and regulations

Although, The Renewable Energy Law and its subsequent amendments are the main pieces of legislation supporting the development of renewable energy, The Electricity Market Law and the Energy Efficiency Law are also major relevant pieces of the Legislation. Together with these Laws, related secondary legislation, regulations and supporting regulations, like Electricity Market Licensing Regulation, set the legal framework for promoting electricity generation from renewables and including main instruments. They offer some advantages like feed-in tariff and purchase obligations, connection priority, reduced license fees, exemption from license and company establishment obligations for the plants with a maximum capacity of 500 kW, reduced land use fees or free land use. (EUMS, 2009). The Turkey's legislation on renewable energy established within the last decade (Table 1).

Date	Legislation
2001 :	Electricity Market Law (No:4628)
2002	Electricity Market Licensing Regulation
2003	Petroleum Market Law (No:5015)
2004	Strategy Paper as Road Map of the Electricity Market Reform & Transition
2005	Law on Utilization of Renewables in Electricity Generation
2007	Energy Efficiency Law (No:5627)
	Amendments to the Law on Utilization of Renewables in Electricity Generation (No:5346)
	Geothermal Law (No:5686)
2008	Significant Amendments to the Electricity Market Law (No:5784)
2009	Strategy Paper on Electricity Market Reform, & Security of Supply
2011	Amended Law on Utilization of Renewables in Electricity Generation (No. 6094)

Table 1. Renewable energy -related legislation (Çetin, 2010).

The Electricity Market Law and Electricity Market Licensing Regulation entered into force in 2001 and 2002, respectively, also set forth a number of pro-renewables provisions (TR, 2001; TR,2002). Turkey enacted its first law specific to renewable energy, the Law on the Utilization of Renewable Energy Sources for the Purpose of Generating Electrical Energy (the Renewable Energy Law) with No. 5346, on 18 May 2005 (TR, 2005). This was a key step for strengthening the country's renewable energy sector. According to this Law, the legal entity holding generation license shall be granted by EMRA with a "Renewable Energy Resource Certificate" (RES Certificate) for the purpose of identification and monitoring of the resource type in purchasing and sale of the electrical energy generated from renewable energy resources in the domestic and international markets. This Law provides feed in tariff until 2011, purchase guarantee. In addition, State territories are permitted on the basis of its sale price, rented, given right of access or usage permission by Ministry of Environment and Forestry or Ministry of Finance. Fifty percent deduction is implemented for permission, rent, and right of access and usage permission in the investment period. Following the enactment of the first Renewable Energy Law, investors showed an increasing interest in Renewables, especially in relation to the generation of electricity through hydro plants and wind farms (Saygın & Çetin, 2010). However, the interest in renewable energy projects was hindered by the lenders' reluctance because of the uncertainty in the purchase guarantees. As a result, the government introduced an important series of amendments in 2007 and 2008(TR, 2007; TR, 2008). The amendment to the Law in May 2007 secured a constant purchase price for all types of renewable sources (Saygın& Cetin, 2010). Together with the Amendment in 2008 entered into force following incentives are offered by Renewable Energy Law to promote renewable energy (TR,2008; EUMS, 2009; Gümüş, 2011; Kolcuoğlu, 2011).

For Renewable Power Plants (PPs) in operation for not longer than 10 years:

- The average electricity wholesale price of the previous year is to be determined by EMRA and limited to €cent 5-5.5/kWh RES,

-    Certificate owners are also granted the right to sell their output at higher rates whenever available in the spot market or via bilateral contracts with eligible customers,
-    The share of renewable output within the retail licensees' portfolio cannot be less than their domestic market share.

During the first 10 years of operation, an 85% deduction is applied to fees related to permission, rent, and right of access and usage permission over the investment and operation period, in the event of the use of the property under the possession of the General Directorate of Forestry or the Treasury.

85% deduction is applied to fees related to investments in the transportation infrastructure and power lines until the connection point to the grid.

Exemption from the special fees charged to contribute to the development of woodland villages, promotion of forestation and erosion mitigation.

Free use of state-owned estates located within the reservoir of Hydroelectric Power Plants holding a RES Certificate.

Following the enactment of the Amended Renewable Energy Law in 2007, investor interest in the renewable energy sector has risen distinctively, and a significant progress has been made. The efforts successfully resulted in an appreciable increase in the share of renewables excluding large- hydro in total especially in the wind and geothermal capacities, as seen from Figure 9.

Despite this hopeful development in hydro, wind and geothermal energy, solar capacity has not developed and clearly needs further promotion. By this aim, a New Amendment to Renewable Energy Law was supposed, therefore, to the National General Assembly on June, 2009. It was suspended until recently since it would create an extra burden on the treasury (Saygın & Çetin, 2010).

Energy source	Generation. [TWh]	Share, %	Capacity, [MW]	Share, %
Natural gas	94.4	48.6	16 345.2	19.4
Domestic Coal	42.2	21.7	8 691.3	19.4
İmported Coal	12.8	6.6	1 921.	4.3
Hydropower	35.9	18.5	14 553.4	32.5
Liquid fuels (Oil)	6.6	3.4	2309.7	5.2
Wind & Geothermal, Biogas	2.2	1.1	961.2	2.1
TOTAL	194.1	100	44 782	100

Table 2. Turkey's installed capacity and power generation by fuel (Çetin, 2010).

By 2009, renewable sources provided 37.8 TWh of electricity, or 19.6% of the total power generation in Turkey. Hydropower accounted for 95 % (35.9 TWh) of this total, wind power for 4 % (1.5 TWh), biomass and geothermal for %1 (0.5 TWh). The Country is the 12th highest share among the 28 IEA countries (IEA, 2009).

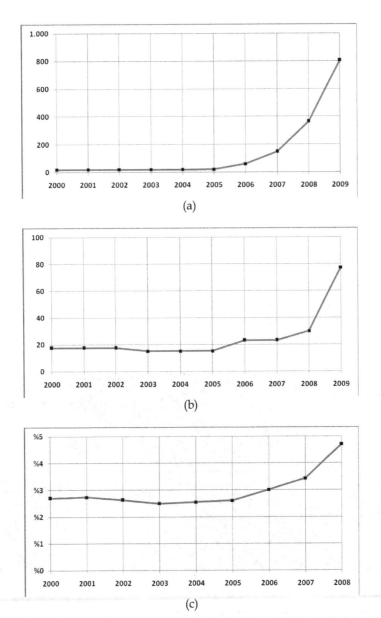

(a)

(b)

(c)

Fig. 9. In Turkey, evaluation of a) installed wind power capacity, b) installed geothermal capacity, c) the share of renewables excluding large- hydro (MENR, 2010).

A total of 601 renewable projects with a capacity of 15500 MW had been licensed by 2009(Saygın & Çetin, 2010) .This number has reached to 645 by 2010. The number of licensed power plants according to energy sources is given in the Figure 10

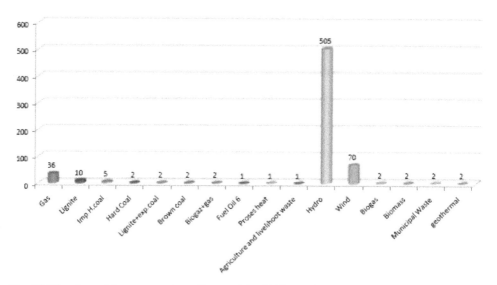

Fig. 10. Number of the power plants licensed by 2010, total 645(Çetin, 2010).

As can be seen from this figure, Hydroelectric, and wind power plant hopefully left behind the others. This implies promotion policy for renewable energy started to be effective. Despite these progressive steps forward, most of the huge renewable potential of Turkey has not been used yet, as can be seen from Figure 11.

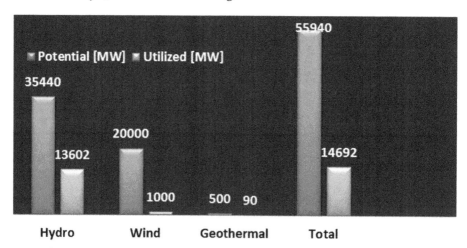

Fig. 11. Utilization Rates of Renewable Energy Potential in Turkey[6]

On 29 December 2010, the Turkish Parliament finally passed this Amendment Law upgrading and differentiating feed-in tariff structure with regards to sources. As seen from the Table 3, the Law guarantees prices of 7.3 US cents per kilowatt-hour for hydroelectric and wind, a price of 10.5 US cents for geothermal energy. a price of 13.3 US cents for solar energy as well as waste products (such as biomass or municipal solid waste-to-energy projects). As can be understood, it has been especially aimed to promote solar energy by this amendment. Under the last Amended Renewable Energy Law, individuals and legal entities generating electricity from renewable resources are able to benefit from this feed-in tariffs, provided that they allocate any excess amount of generated electricity to the national electricity distribution system. This will be applicable for a period of 10 years for generation license holders who are subject to RES support mechanism and have commenced/will commence operations between 18 May 2005 and 31 December 2015 (TR, 2010).

Plant Type	Purchase price for produced electricity USD cents/kWh
Hydro	7.3
Wind	7.3
Geothermal	10.5
Biomass	13.3
Solar	13.3

Table 3. Feed in Tariff for Renewables (TR, 2010).

The Law also offers add payment for components made in Turkey. If the mechanical and electro-mechanical equipment used in renewable energy facilities that have started operation before 31 December 2015 are manufactured in Turkey, an additional incentive of 0.4-2.4 USD/cent for five years will be provided. In addition EMRA will give priority to facilities generating energy from renewable resources in evaluating generation license applications. In addition, The Amendment Renewable Law offers Incentives through the Pooling of Payments: It envisages a pool managed by the Market Financial Settlement Centre (MFSC) whereby the electricity suppliers will make the payment of the renewable energy and the renewable energy generators will collect their fees (TR, 2011; Kolcuoğlu, 2011).

Other incentives provided in the legislation are as follows:

- Legal entities applying for a license for the construction of facilities based on domestic natural resources and RES shall only pay 1 per cent of the total licensing fee.
- Generation facilities based on renewable and domestic energy resources shall not pay annual license fees for the first eight years following the facility completion date inserted in their respective license.
- TEIAS, the State Owned Transmission Company, and distribution licensees shall give priority to the system connection of generation facilities based on domestic natural resources and renewable resources.

If the price of electricity generated at facilities based on RES is equal to or lower than the sales price of TETAS, the state owned wholesale company, and if there is no cheaper alternative, the retail licensees are obliged to purchase the electricity generated at facilities based on RES for the purposes of resale to non-eligible consumers.

With the passing of the amendments to the Turkish Renewable Energy Law in December 2010, more progressive development in renewable energy sector can be expected. Although none of the legislation fully met the needs of renewable energy investors – until now and the incentives provided for renewable energy investments are criticized by the investors since the feed-in tariffs are lower than expected, a certain progress have been made. The efforts made so far are hopeful; however it seems there are more steps to be taken.

4. Place of renewables in Turkey's near and long term energy strategy

Presently, Turkey is at the stage of setting targets of renewable energy development. The Higher Board of Planning adopted the "Electric Energy Market and Supply Security Strategy Paper for long term target in May 2009 (MENR, 2009). In this Strategy Paper, the long term primary target is determined as "to ensure that the share of renewable resources in electricity generation is increased up to at least 30% by 2023 (MENR, 2010)

This strategy document published as a general road map of increasing the share of renewable energy in electricity generation.

Within the framework of the Strategy Paper, long term works will take into consideration the following targets, by 2023(MENR, 2010):

- It will be ensured that technically and economically available hydro-electric potential is entirely put to use,
- It is targeted to increase installed wind energy power to 20, 000 MW,
- Turkey's geothermal potential of 600 MW, which is presently established as suitable for electric production, is entirely commissioned by 2023.
- It is targeted to generalize the use of solar energy for generating energy, ensuring maximum utilization of country potential. Regarding the use of solar energy for electricity generation, technological advances will be closely followed and implemented.
- Preparation of production Plans will take into account potential changes in utilization potentials of other renewable energy resources based on technological and legislative developments and in case of increases utilization of such resources, share of fossil fuels and particularly of imported resource, will be reduced accordingly.

For determining near-term targets, Ministry of Energy and Natural Resources prepared a Strategic Plan covering the period between 2010 and 2014 (MENR, 2010), which is given in Table 4. For the purpose of increasing the energy supply security, the resources, routes and technologies will be diversified. Beside energy efficiency, increasing use of renewables and the integration of nuclear energy into energy mix are the two main components of new energy policies of Turkey. According to the Plan's aim of providing diversification of the energy supply, the maximum use of the domestic and renewable resources in the production of electricity energy and the initiation of the construction of the nuclear plant have been targeted.

It should be emphasized that both of this long term and near term strategy plans are not, on their own, legally binding. It is, however, expected that their provisions will be incorporated into future regulations and legislation.

THE REPUBLIC OF TURKEY MINISTRY OF ENERGY AND NATURAL RESOURCES STRATEGIC PLAN (2010-2014)

Aims For Energy Supply Security

1. Providing Diversity in Resources by Giving Priority to the Domestic Resources
    Target 1.1 Within the period of the Plan, the domestic oil, natural gas and coal exploration works will be increased.
    Target 1.2 The domestic coal thermal plants of 3,500 Mega Watt (MW) will be completed by 2013.
    Target 1.3 By the year 2014, the construction of nuclear plant will start.

2. Increasing the share of the renewable energy resources within the energy supply
    Target 2.1 The construction of The hydroelectricity plants of 5,000 MW, will be completed by 2013.
    Target 2.2 The wind plant installed capacity, which has been 802,8 MW as of 2009 will be increased up to 10,000 MW by 2015.
    Target 2.3 The installed capacity for the geothermal plant of 77,2 MW in 2009, will be increased up to 300 MW until 2015.

3. Increasing Energy Efficiency
    Target 3.1 Within framework of the energy efficiency studies, 10 % reduction in energy consumption will be secured by 2015 in comparison to 2008.
    Target 3.2 The completion of the maintenance, rehabilitation and modernization studies conducted for increasing the efficiency and production capacity through the use of new technologies in the existing state owned electricity production plants by the end of 2014 will be secured.

4. Making the free market conditions operate fully and providing for the improvement of the investment environment
    Target 4.1 By the year 2014, the targeted privatizations in the electricity sector will be completed.
    Target 4.2 By the year 2015, the formation of the market structure that works as based on competition will be secured.
    Target 4.3 By the year 2015, the formation of the market structure that works as based on competition will be secured in the natural gas sector.

5. Providing the diversity of resources in the area of oil and natural gas and taking the measures for reducing the risks due to importation
    Target 5.1 By the year 2015, the foreign crude oil and natural gas production will be redoubled in comparison to the production amounts in 2008.
    Target 5.2 The existing natural gas storage capacity which is 2,1 billion m^3 in 2009 will be redoubled by 2015.
    Target 5.3 In natural gas importation, by the year 2015, we will decrease the share of the country from which the highest amount of importation is made and the diversity of source countries will be provided.
    Target 5.4 The sustainability of the storage of the national oil stocks at a secure level will be provided.

Table 4. Turkey's near term Strategic Plan for the Period 2010-2014(MENR, 2010).

5. Renewable energy projects

Turkey is within some projects for reaching its aims related to renewable energy. The most important one of them is The Private Sector Renewable Energy and Energy Efficiency Project. The Project aims to help increase privately owned and operated energy production from indigenous renewable sources within the market-based framework of the Turkish Electricity Market Law, thereby helping to enhance energy efficiency and curb greenhouse gas emissions as a result. The project is the first to use resources from the newly established Clean Technology Fund (CTF) — a new US$5.2 billion multilateral fund managed by the World Bank and administered through the World Bank Group and other multilateral development banks. World Bank and CTF financing for the project will include two loans, one from the IBRD and one from the CTF, each to two Turkish development banks: Türkiye Sinai Kalkınma Bankası (TSKB) the Turkish Industrial Development Bank (private) and Türkiye Kalkınma Bankası (TKB) the Turkish Development Bank (government) — for credit line financing of renewable energy and energy-efficiency investments(WB, 2011).

Another important Project is Renewable Energy Networks (RENET) between Turkish and European Universities. The project allows the combination of best available technology information in the field of renewable energy and best practices transfer in higher education and science and Technology co-operation.

The specific objectives of the RENET project are to (RENET, 2011):

i.   enhance the awareness and understanding of academic staff as well as of policy-makers and stakeholders of the political, economical and social frameworks relating renewable energy;
ii.   promote the Information and Technology Transfer for renewable energies;
iii. increase the capacity of the academic community to participate in Turkish-European co-operation projects, and
iv.   foster the strategic partnership for implementing renewable energies in Turkey and the EU.

It is expected that RENET will provide a relevant contribution not only to the Millennium Development Goals (MDGs) on ensuring environmental sustainability and developing global partnerships for development. Moreover, it is also in line with the strategic objectives of the Lisbon-Gothenburg Strategy of the European Union(RENET, 2011).

6. Conclusions

Turkish energy market is currently the scene of important changes. The Country established a new national energy plan based on diversification of supplies, the start of nuclear energy production and development of renewable energy and energy efficiency. It seems to move towards a low carbon energy sector: In this context, it has three main strategies: increasing energy efficiency and renewable energy use in addition increasing natural gas use. This implies Turkey is stepping up its engagement on climate change internationally and nationally.

It has taken some important steps to promote the use of renewable energy resources. The Country's renewable energy policies are in a consistency with EU Energy Policies and global trends like energy efficiency and market liberalization policies. It is clearly seen the country is heading in the right direction. However, its deliberate policy has caused relatively slow progress in the realization of its aims. This may cause Turkey to miss the technological leapfrogging (Saygın, 2006) opportunity presented by emerging paradigm for developing countries. In this context, Turkey has remained behind other leading developing countries like China and India.

7. References

Çalıkoğlu, E. (2007).Energy Efficiency in Turkey, *TAIEX Workshop 25625 on Demand Side Management in Energy Efficiency*, 22-23 November 2007, Turkey, Available from http://www.eie.gov.tr/duyurular/EV/TAIEX/ErdalCalikoglu_TAIEXWorkshop2 5625onDSM_221107.pdf.

Çetin, H. (2010).Turkish Energy Sector ,*UNECE-e8-EBRD-WEC Fostering Investment in Electricity Generation in Central and Eastern Europe and Central Asia*, November 22-24, 2010, Geneva, Switzerland, Available from http://www.google.com.tr/webhp?hl=tr#hl=tr&source=hp&q=Turkish+Energy+ Sector+%2CUNECE-e8-EBRD-.

EUMS (Energy, Utilities & Mining Sector), (2009) Renewables Report On the sunny side of the street* Opportunities and challenges in the Turkish renewable energy market Industries, Available from www.boell-meo.org/downloads/Renewable_Energy_Turkey.pdf.

Gümüş, S.,(2011) ,Turkey: Incentives For Renewable Energy, Energy & Natural Resources, on 28 April 2011. Available from http://www.mondaq.com/x/130778/Renewables/Incentives+For+Renewable+ Energy.

IEA(2009), Energy Policies of IEA Countries, Turkey Review 2009, Paris, France.

Kolcuoglu D.(2011). Incentives under the Long-Awaited Renewable Energy Law – Turkey, HG.org, Available from http://www.hg.org/article.asp?id=20931.

MENR (2009), Turkey's Electric Energy Market and Supply Security Strategy Paper with Res. No. 2009/11, dated 18.09.2009, Available from http://www.enerji.gov.tr/yayinlar_raporlar/Arz_Guvenligi_Strateji_Belgesi.pdf.

MENR (2010). The Republic of Turkey Ministry of Energy and Natural Resources Strategy Plan 2010-2014, Ankara, Turkey.

Observatoire Mediterraneen de l'Energie (OME),(2008) Mediterranean Energy Perspectives 2008.

Özçaldıran, K (2010). A profile of Turkish Energy Sector and related greenhouse gas emission", *The Symposium on Global Energy Future,* 1-5 October, 2010, St. Louis, USA.

Saygın, H. (2006). Technological leapfrogging in energy in developing countries (in Turkish)", *Enerji,* Vol.11, No.1 , pp. 27.

Saygın, H. and Cetin F. (2010). New energy paradigm and renewable energy:Turkey's vision, *Insight Turkey,* Vol. 12, No. , pp. 107-128.

TR (2001). The Electricity Market Law No. 4628, *Official Gazette,* No. 24335, March 2001.

TR (Republic of Turkey), (2005). Law on Utilization of Renewable Energy Sources for the Purpose of Generating Electrical Energy, *Official Gazette*, with No. 24335, dated 10.05.2005 (Law No.: 5346).

TR, (2007). The Law Amending the Law on Utilization of Renewable Energy Resources in Electricity Generation , *Official Gazette* ,with No. 26510, dated 02.05.2007.

TR, (2008).The Law Amending the Law on Utilization of Renewable Energy Resources in Electricity Generation Law on Utilization of Renewable Energy Sources for the Purpose of Generation Electricity, *Official Gazette*, No. 270522, dated 03.12.2008.

TR, (2011). The Law Amending the Law on Utilization of Renewable Energy Resources in Electricity Generation (Law No: 6094), *Official Gazette dated 8 January 2011 and numbered 27809.*

WB (World Bank), (2011). Turkey: Private Sector Renewable Energy and Energy Efficiency Project, Available from
http://www.worldbank.org.tr/WBSITE/EXTERNAL/COUNTRIES/ECAEXT/TU
RKEYEXTN/0,,contentMDK:22244557~menuPK:361754~pagePK:2865066~piPK:28
65079~theSitePK:361712,00.html.

RENET,(2011). Renewable Energy Networks between Turkish and European Universities , Available from http://www.renet-project.eu/.

4

Smart Dispatch and Demand Forecasting for Large Grid Operations with Integrated Renewable Resources

Kwok W. Cheung
Alstom Grid Inc.
USA

1. Introduction

The restructured electric power industry has brought new challenges and concerns for the secured operation of stressed power systems. As renewable energy resources, distributed generation, and demand response become significant portions of overall generation resource mix, smarter or more intelligent system dispatch technology is needed to cope with new categories of uncertainty associated with those new energy resources. The need for a new dispatch system to better handle the uncertainty introduced by the increasing number of new energy resources becomes more and more inevitable.

In North America, almost all Regional Transmission Organizations (RTO) such as PJM, Midwest ISO, ISO New England, California ISO or ERCOT, are fundamentally reliant on wholesale market mechanism to optimally dispatch energy and ancillary services of generation resources to reliably serve the load in large geographical regions. Traditionally, the real-time dispatch problem is solved as a linear programming or a mixed integer programming problem assuming absolute certainty of system input parameters and there is very little account of system robustness other than classical system reserve modeling. The next generation of dispatch system is being designed to provide dispatchers with the capability to manage uncertainty of power systems more explicitly.

The uncertainty of generation requirements for maintaining system balancing has been growing significantly due to the penetration of renewable energy resources such as wind power. To deal with such uncertainty, RTO's require not only more accurate demand forecasting for longer-term prediction beyond real-time, but also demand forecasting with confidence intervals.

This chapter addresses the challenges of smart grid from a generation dispatch perspective. Various aspects of integration of renewable resources to power grids will be discussed. The framework of Smart Dispatch will be proposed. This chapter highlights some advanced demand forecasting techniques such as wavelet transform and composite forecasting for more accurate demand forecasting that takes renewable forecasting into consideration. A new dispatch system to provide system operators with look-ahead capability and robust dispatch solution to cope with uncertain intermittent resources is presented.

2. Challenges of smart grid

In recent years, energy systems whether in developed or emerging economies are undergoing changes due to the emphasis of renewable resources. This is leading to a profound transition from the current centralized infrastructure towards the massive introduction of distributed generation, responsive/controllable demand and active network management throughout the smart grid ecosystem as shown in Figure 1. Unlike conventional generation resources, outputs of many of renewable resources do not follow traditional generation/load correlation but have strong dependencies on weather conditions, which from a system prospective are posing new challenges associated with the monitoring and controllability of the demand-supply balance. As distributed generations, demand response and renewable energy resources become significant portions of overall system installed capacity, a smarter dispatch system for generation resources is required to cope with the new uncertainties being introduced by the new resources.

One method to cope with uncertainties is to create a better predictive model (Cheung et al., 2010, 2009). This includes better modeling of transmission constraints, better modeling of resource characteristics including capacity limits and ramp rates, more accurate demand forecasting and external transaction schedule forecasting that ultimately result in a more accurate prediction of generation pattern and system conditions. Another method to cope with uncertainties is to address the robustness of dispatch solutions (Rios-Zalapa et al., 2010). Optimality or even feasibility of dispatch solutions could be very sensitive to system uncertainties. Reserve requirements and "n-1" contingency analysis are traditional ways to ensure certain robustness of a given system. Scenario-based (Monte-Carlo) simulation is another common technique for assessing economic or reliability impact with respect to uncertainties such as renewable energy forecast. These methods and techniques are necessary as the industry integrates renewable energy resources into the power grid.

2.1 Renewable energy grid integration

Like any other form of generation, renewable resources such as wind or solar power will have an impact on power system reserves and will also contribute to a reduction in fuel usage and emissions. In particular, the impact of wind power not only depends on the wind power penetration level, but also on the power system size, geographical area, generation capacity mix, the degree of interconnection to neighboring systems and load variations.

Some of the major challenges of renewable energy integration need to be addressed in the following main areas:

- Design and operation of the power system
- Grid infrastructure
- Connection requirements for renewable power plants
- System adequacy and the security of supply
- Electricity market design

With increasing penetration and reliance on renewable resources have come heightened operational concerns over maintaining system balance. Ancillary services, such as operating reserves, imbalance energy, and frequency regulation, are necessary to support renewable energy integration, particularly the integration of intermittent resources (Chuang & Schwaegerl, 2009). Without supporting ancillary services, increased risk to system imbalance is introduced by the uncertainty of renewable generation availability, especially in systems with significant penetration of resources powered by intermittent supply, such as wind and solar.

For the purposes of balancing, the qualities of wind energy must be analyzed in a directly comparable way to that adopted for conventional plants. Balancing solutions involve mostly existing conventional generation units (thermal and hydro). In future developments of power systems, increased flexibility should be encouraged as a major design principle (flexible generation, demand side management, interconnections, storage etc.), in order to manage the increased variability induced by renewable resources. Market design issues such as gate-closure times should be reduced for variable output technologies. The real-time or balance market rules must be adjusted to improve accuracy of forecasts and enable temporal and spatial aggregation of wind power output forecasts. Curtailment of wind power production should be managed according to least-cost principles from an overall system point of view.

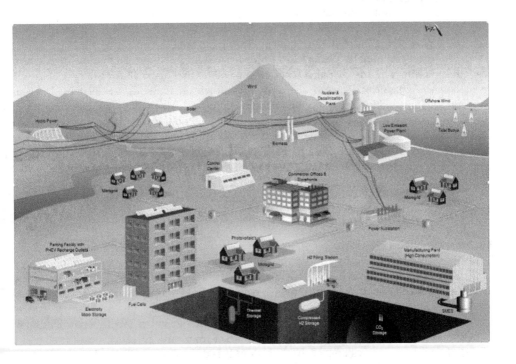

Fig. 1. Smart Grid Ecosystem

3. Smart dispatch of generation resources

Smart dispatch (SD) represents a new era of economic dispatch. In general, economic dispatch is about the operation of generation facilities to produce energy at the lowest cost to reliably serve consumers, recognizing any operational limits of generation and transmission facilities. The problem of economic dispatch and its solutions have evolved over the years.

3.1 Evolution of economic dispatch

The evolution timeline of economic dispatch could be divided into the following three major periods:
1. Classical dispatch [1970's – 1990's] (Wood & Wollenberg, 1996)
2. Market-based dispatch [1990's – 2010's] (Schweppe et al., 1998; Ma et al., 1999; Chow et al., 2005)
3. Smart dispatch [2010's –] (Cheung et al., 2009)

3.1.1 Classical dispatch

Since the birth of control center's energy management system, classical dispatch monitors load, generation and interchange (imports/exports) to ensure balance of supply and demand. It also maintains system frequency during dispatch according to some regulatory standards, using Automatic Generation Control (AGC) to change generation dispatch as needed. It monitors hourly dispatch schedules to ensure that dispatch for the next hour will be in balance. Classical dispatch also monitors flows on transmission system. It keeps transmission flows within reliability limits, keeps voltage levels within reliability ranges and takes corrective action, when needed, by limiting new power flow schedules, curtailing existing power flow schedules, changing the dispatch or shedding load. The latter set of monitoring and control functions is typically performed by the transmission operator. Traditionally, generation scheduling/dispatch and grid security are separate independent tasks within control centers. Other than some ad hoc analysis, classical dispatch typical only addresses the real-time condition without much consideration of scenarios in the past or the future.

3.1.2 Market-based dispatch

Ensuring reliability of the physical power system is no longer the only responsibility for the RTO/ISOs. A lot of the RTOs/ISOs are also responsible for operating wholesale electricity markets. An electricity market in which the ISO or RTO functions both as the "system operator" for reliability coordination and the "market operator" for establishing market prices allows commercial freedom and centralized economic and reliability coordination to co-exist harmoniously (Figure 2). To facilitate market transparency and to ensure reliability of the physical power system, an optimization-based framework is used to provide an Taking advantage of the mathematical rigor contained in formal optimization methodology, the rules are likely to be more consistent, and thus more defensible against challenges that effective context for defining comprehensive rules for scheduling, pricing, and dispatching. invariably arise in any market.

Congestion management via the mechanism of locational marginal pricing (LMP) becomes an integral part of design of many wholesale electricity markets throughout the world and security-constrained economic dispatch (SCED) becomes a critical application to ensure the transmission constraints are respected while generation resources are being dispatched economically. The other important aspect of market-based dispatch is the size of the dispatch system. A typical system like PJM or Midwest ISO is usually more than 100GW of installed capacity. Advances in mathematical algorithms and computer technology really make the near real-time dispatch and commitment decisions a reality.

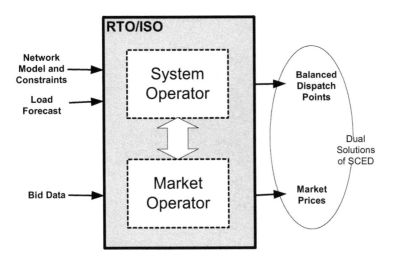

Fig. 2. Dual functions of RTO/ISO and dual solutions of SCED

3.1.3 Smart dispatch

Smart dispatch (SD) is envisioned to be the next generation of resource dispatch solution particularly designed for operating in the smart grid environment (Cheung et al., 2009). The "smartness" of this new era of dispatch is to be able to manage highly distributed and active generation/demand resources in a direct or indirect manner. With the introduction of distributed energy resources such as renewable generations, PHEVs (Plug-in Hybrid Electric Vehicles) and demand response, the power grid will need to face the extra challenges in the following areas:

- Energy balancing
- Reliability assessment
- Renewable generation forecasting
- Demand forecasting
- Ancillary services procurement
- Distributed energy resource modeling

A lot of the new challenges are due to the uncertainties associated with the new resources/devices that will ultimately impact both system reliability and power economics. When compared to the classical dispatch which only deals with a particular scenario for a single time point, smart dispatch addresses a spectrum of scenarios for a specified time period (Figure 3). Thus the expansion in time and scenarios for SD makes the problem of SD itself pretty challenging from both a computational perspective and a user interface perspective. For example, effective presentation of multi-dimensional data to help system operators better visualize the system is very important. Beside a forward-looking view for system operators, SD should also allow after-the-fact analysis. System analysts should be able to analyze historical data systematically and efficiently, establish dispatch performance measures, perform root-cause analysis and evaluate corrective actions, if necessary. SD will become an evolving platform to allow RTOs/ISOs to make sound dispatch decisions.

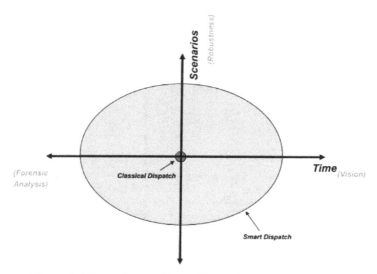

Fig. 3. Time and Scenario Dimensions in Smart Dispatch

3.2 Framework of smart dispatch

The objective of this section is to reveal the proposed framework of Smart Dispatch. The framework outlines the basic core SD functions for RTOs/ISOs operating in the smart grid environment. Some of the functional highlights and differentiations from classical dispatch are:

- Extension for price-based, distributed, less predictable resources
- Active, dynamic demand
- Modeling parameter adaptation
- Congestion management with security constrained optimization
- Continuum from forward scheduling to real-time dispatch
- Extension for dynamic, multi-island operation in emergency & restoration
- After-the-fact analysis for root-cause impacts and process re-engineering

One major core functions of Smart Dispatch is called Generation Control Application (GCA) which aims at enhancing operators' decision making process under changing system conditions (load, generation, interchanges, transmission constraints, etc.) in near real-time. GCA is composed of several distinct elements (Figure 4):

- Multi-stage Resource Scheduling Process (SKED 1,2&3)
- Comprehensive Operating Plan (COP)
- Adaptive Model Management

The multi-stage resource scheduling (SKED) process is security constrained unit commitment and economic dispatch sequences with different look-ahead periods (e.g. 6 hours, 2 hours and 20 minutes) updating resource schedules at different cycle frequencies (e.g. 5min, 15min or hourly). The results of each stage form progressively refined regions that guide the dispatching decision space of the subsequent stages. Various SKED cycles are coordinated through the so-called Comprehensive Operating Plan (COP).

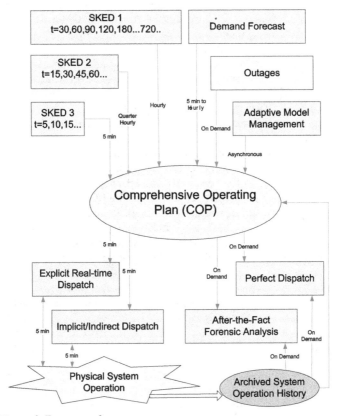

Fig. 4. Smart Dispatch Framework

COP is a central repository of various kinds of scheduling data to and from a certain class of power system applications. COP presents a comprehensive, synchronized and more harmonized view of scheduling data to various applications related to power system operations. The class of scheduling data of interest includes the followings:

- Resource (renewable/non-renewable) MW schedule
- Demand forecast
- Outage schedule
- Transaction and interchange schedule
- Transmission constraint limit schedule
- Reserve and regulation requirement schedule
- Resource characteristics schedule

COP also contains comprehensive summary information. Summary information could be rollups from a raw data at a lower level (e.g. resource level) according to some pre-defined system structures.

Adaptive Model Management as shown in Figure 4 consists of two parts: Advanced Constraint Modeling (ACM) and Adaptive Generator Modeling (AGM). ACM will use intelligent methods to preprocess transmission constraints based on historical and current network conditions, load forecasts, and other key parameters. It should also have ability to achieve smoother transmission constraint binding in time. AGM will provide other GCA components with information related to specific generator operational characteristics and performances. The resource "profiles" may contain parameters such as ramp rate, operating bands, predicted response per MW of requested change, high and low operating limits, etc.

Another major core functions of Smart Dispatch is After-the Fact Analysis (AFA). AFA aims at providing a framework to conduct forensic analysis. AFA is a decision-support tool to:

a.   Identify root cause impacts and process re-engineering.
b.   Systematically analyze dispatch results based on comparison of actual dispatches with idealized scenarios.
c.   Provide quantitative and qualitative measures for financial, physical or security impacts on system dispatch due to system events and/or conditions.

One special use case of AFA is the so-called "Perfect Dispatch" (PD). The idea of PD was originated by PJM (Gisin et al., 2010). PD calculates the hypothetical least bid production cost commitment and dispatch, achievable only if all system conditions were known and controllable. PD could then be used to establish an objective measure of RTO/TSO's performance (mean of % savings, variance of % savings) in dispatching the system in the most efficient manner possible by evaluating the potential production cost saving derived from the PD solutions.

Demand forecast is a very crucial input to GCA. The accuracy of it very much impacts market efficiency and system reliability. The following is devoted to discuss some recent advances in techniques of demand forecasting.

4. Demand forecast

Demand or load forecasting is very essential for reliable power system operations and market system operations. It determines the amount of system load against which real-time dispatch and day-ahead scheduling functions need to balance in different time horizon. Demand forecasting typically provides forecasts for three different time frames:

1.   Short-Term (STLF): Next 60-120 minutes by 5-minute increments.
2.   Mid-Term (MTLF): Next n days (n can be any value from 3-31), in intervals of one hour or less (e.g., 60, 30, 20, 15 minute intervals).
3.   Long-Term (LTLF): Next n years (n can be any value from 2-10), broken into one month increments. The LTLF forecast is provided for three scenarios (pessimistic growth, expected growth, and optimistic growth).

Demand forecasting play an increasingly important role in the restructured electricity market and smart grid environment due to its impacts on market prices and market participants' bidding behavior. In general, demand forecasting is a challenging subject in view of complicated features of load and effective data gathering. With Demand Response being one of the few near-term options for large-scale reduction of greenhouse gases, and fits strategically with the drive toward clean energy technology such as wind and solar, advanced demand forecasting should effectively take the demand response features/characteristics and the uncertainty of interimttent renewable generation into account.

Many load forecasting techniques including extrapolations, autoregressive model, similar day methods, fuzzy logic, Kalman filters and artificial neural networks. The rest of section will focus on the discussion of STLF which is a key input to near real-time generation dispatch in market and system operations.

4.1 The uncertainty of demand forecast

The uncertainty for demand forecast is one of the most critical factors influencing the uncertainty of generation requirements for system balancing (DOE, 2010). It is important to note that wind generation has fairly strong positive correlation with electrical load in many ways more than traditional dispatchable generation. As a result, it is viable to treat wind generation as a negative load and incorporate its uncertainty analysis as part of the uncertainty of demand forecast assuming transmission congestion is not an issue. Hence, the concept of net demand has been employed in wind integration studies to assess the impact of load and wind generation variability on the power system operations. Typically, the net demand has been defined as the following:

Net demand = Total electrical load – Renewable generation + Net interchange

One practical approach can be used for the uncertainty modeling of demand forecast is distribution fitting. Basically probability distributions are based on assumptions about a specific standard form of random variables. Based on the standard distributions (e.g. normal) and selected set of its parameters (e.g. mean μ, standard deviation σ), they assign probability to the event that the random variable x takes on a specific, discrete value, or falls within a specified range of continuous values. An example of the probability density function $PDF(x)$ (Meyer, 1970) of demand forecast is presented in Figure 5a. The cumulative distribution function $CDF(x)$ can then be defined as:

$$CDF(x) = \int_{-\infty}^{x} PDF(s)ds \qquad (1)$$

A confidence interval (CI) is a particular kind of interval estimate of a population parameter such that the random parameter is expected to lie within a specific level of confidence. A confidence interval in general is used to indicate the reliability of an estimate and how likely the interval to contain the parameter is determined by the confidence level (CL). The CL of confidence interval $[Dl, Dh]$ for demand forecast can be defined as:

$$CL(Dl \le x \le Dh) = \{CDF(Dh) - CDF(Dl)\} \times 100\% \qquad (2)$$

Increasing the desired confidence level will widen the confidence interval being controlled by parameters $k1$ and $k2$ as shown in Figure 5. It is obvious that the size of uncertainty ranges depends on the look-ahead time. In general for longer look-ahead periods, the uncertainty range becomes larger. Figure 6 illustrates the time-dependent nature of confidence intervals – cone of uncertainty for demand forecast.

4.2 Artificial neural network with wavelet transform

In the era of smart grid, the generation and load patterns, and more importantly, the way people use electricity, will be fundamentally changed. With intermittent renewable generation, advanced metering infrastructure, dynamic pricing, intelligent appliances and HVAC equipment, micro grids, and hybrid plug-in vehicles, etc., load forecasting with uncertain factors in the future will be quite different from today. Therefore, effective STLF are highly needed to consider the effects of smart grid.

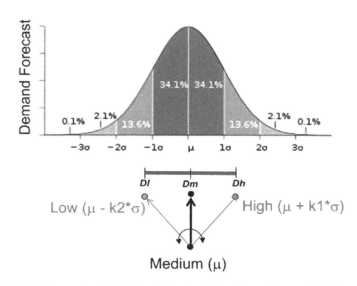

Fig. 5. Probabilistic Uncertainty Model and Desired Confidence Interval for Demand Forecast

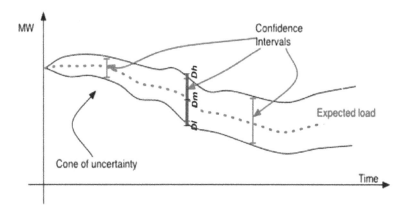

Fig. 6. Confidence Intervals for Demand Forecast

Based on frequency domain analysis, the 5-minute load data have multiple frequency components. They can be illustrated via power spectrum magnitude. Figure 7 shows a typical power spectrum of actual load of a regional transmission organization. Note that the power density spectrum can be divided into multiple frequency ranges.

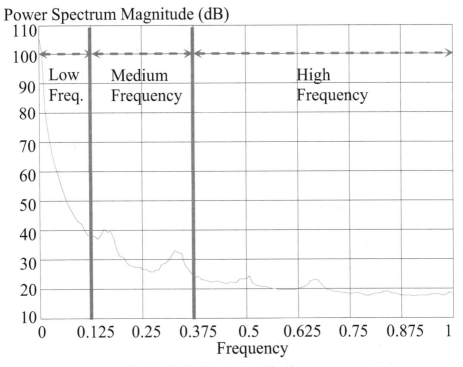

Fig. 7. A power spectrum density for 5-minute actual load

Neural networks have been widely used for load forecasting. They have been used for load forecasting in era of smart grid (Amjady et al., 2010; Zhang et al., 2010). In particular, Chen et al. have presented the method of similar day-based wavelet neural network approach (Chen, et al., 2010). The key idea there was to select "similar day load" as the input load, use wavelet decomposition to decompose the load into multiple components at different frequencies, apply separate neural networks to capture the features of the forecast load at individual frequencies, and then combine the results of the multiple neural networks to form the final forecast (see Figure 9). In general, these methods used general neural networks which adopted multilayer perception with the back-propagation training. There are many wavelet decomposition techniques. Some recent techniques applying to load forecasting are:

- Daubechies 4 wavlet (Chen et al., 2010)
- Multiple-level wavelet (Guan et al., 2010)
- Dual-tree M-band wavelet (Guan et al., 2011)

The Daubechies 4 (D4) wavelet is part of the family of orthogonal wavelets defining a discrete wavelet transform that decomposes a series into a high frequency series and a low frequency series. Multiple-level wavelet basically repeatedly applies D4 wavelet decomposition to the low frequency component of its previous decomposition as shown in Figure 8. Unlike D4 wavelet, Dual-tree M-band wavelet can selectively decompose a series into specified frequency ranges which could be key design parameters for more effective decomposition.

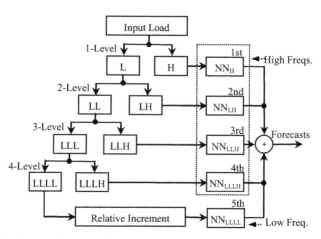

Fig. 8. Multiple-level Wavelet Neural Network

In general, each (Neural Network) NN as shown in Figure 8 could be implemented as a feed-forward neural network being described by the following equation:

$$L_{t+l} = f(t, L_t, L_{t-1}, \cdots, L_{t-n}) + \varepsilon_{t+1} , \tag{3}$$

where t is time of day, l is the time lead of the forecast, L_t is the load component or relative increment of the load component at time t and ε_{t+1} represents a random load component. The nonlinear function f is used to represent the nonlinear characteristics of a given neural network.

4.3 Neural networks trained by hybrid kalman filters

Since back-propagation algorithm is a first-order gradient-based learning algorithm, neural networks trained by such algorithm cannot produce the covariance matrix to construct dynamic confidence interval for the load forecasting. Replacing back-propagation learning, wavelet neural networks trained by hybrid Kalman filters are developed to forecast the load of next hour in five-minute steps with small estimated confidence intervals.

If the NN input-output function was nearly linear, through linearization, NNs can be trained with the extended Kalman filter (EKFNN) by treating weight as state (Singhal & Wu, 1989). To speed up the computation, EKF was extended to the decoupled EKF by ignoring the interdependence of mutually exclusive groups of weights (Puskorius &Feldkamp, 1991). The numerical stability and accuracy of decoupled EKF was further improved by U-D factorization (Zhang & Luh, 2005). If the NN input-output function was highly nonlinear, EKFNN may not be good since mean and covariance were propagated via linearization of the underlying non-linear model. Unscented Kalman filter (Julier et al, 1995) was a potential method, and NNs trained by unscented Kalman filter (UKFNN) showed a superior performance. EKFNN was used to capture the feature of low frequency, and UKFNNs for those of higher frequency. Results are combined to form the final forecast.

To capture the near linear relation between the input and output of the NN for the low component, a neural network trained by EKF is developed through treating the NN weight as the state and desired output as the observation. The input-output observations for the model can be represented by the set $\{u(t), z(t+1)\}$, where $u(t) = \{u_1, ..., u_{nu}\}^T$ is a $nu \times 1$ input vector, and $z(t+1) = z(t+1|t) = \{z_1, ...,z_{nz}\}^T$ is $nz \times 1$ a output vector. Correspondingly, $\hat{z}(t+1)\left(= \hat{z}(t+1|t)\right)$ represents the estimation for measurement $z(t+1)$. The formulation of training NN through EKF (Zhang and Luh, 2005; Guan et al., 2010) can be described by state and measurement functions:

$$w(t+1)=w(t)+\varepsilon(t), \tag{4a}$$

$$z(t+1)=h\big(u(t),w(t+1)\big)+v(t+1), \tag{4b}$$

where $h(\bullet)$ is the input-output function of the network, $\varepsilon(t)$ and $v(t)$ are the process and measurement noises. The former is assumed to be white Gaussian noised with a zero mean and a covariance matrix $Q(t)$, whereas the latter is assumed to have a student t-distribution with covariance matrix $R(t)$. The weight vector $w(t)$ has a dimension $n_w \times 1$ and n_w is determined by numbers of inputs, hidden neurons and outputs:

$$n_w = (n_x + 1) \times n_h + (n_h + 1) \times n_z. \tag{5}$$

Using the input vector $u(t)$, weight vector $w(t)$ and output vector $\hat{z}(t+1)$, EKFNN are derived. Key steps of derivation for EKF (Bar-Shalom et al. 2001) are summarized:

$$\hat{w}(t+1|t) = w(t|t), \tag{6}$$

$$P(t+1|t) = P(t|t) + Q(t), \tag{7}$$

$$\hat{z}(t+1|t) = h\big(u(t),\hat{w}(t+1|t)\big), \tag{8}$$

$$S(t+1) = H(t+1)\cdot P(t+1|t)\cdot H(t+1)^T + R(t+1), \tag{9}$$

$$\text{where} \quad H(t+1) = \big(\partial h(u,w)/\partial w\big)\Big|_{\substack{u=u(t)\\w=\hat{w}(t+1|t)}}, \tag{10}$$

$$K(t+1) = P(t+1|t)\cdot H(t+1)^T \cdot S(t+1)^{-1}, \tag{11}$$

$$\hat{w}(t+1|t+1) = \hat{w}(t+1|t) + K(t+1)\cdot\big(z(t+1) - \hat{z}(t+1|t)\big), \tag{12}$$

$$P(t+1|t+1) = P(t+1|t) - K(t+1)\cdot H(t+1)\cdot P(t+1|t). \tag{13}$$

where $H(t+1)$ is the partial derivative of $h(\bullet)$ with respect to $w(t)$ with dimension $n_z \times n_w$, $K(t+1)$ is the Kalman gain, $P(t+1|t)$ is the prior weight covariance matrix and is updated to posterior weight covariance matrix $P(t+1|t+1)$ based on the Bayesian formula, and $S(t+1)$ is the measurement covariance matrix.

Let us denote $\hat{z}_L(t+1|t) = \hat{z}(t+1|t)$ and $\hat{\sigma}_L^2(t+1) = S(t+1)\cdot I_{ny}\cdot(1\cdots1)^T$, where I_{ny} is the unit matrix, $(1\cdots1)^T$ is a vector with length of ny, $\hat{\sigma}_L^2(t+1)$ is the variance vector consists of the diagonal elements of S(t+1). $\hat{z}_L(t+1|t)$ and $\hat{\sigma}_L^2(t+1)$ representing the low frequency component of prediction and variance, respectively, will be used for the final load prediction and confidence interval estimation. Corresponding medium frequency components of $\hat{z}_M(t+1|t)$ and $\hat{\sigma}_M^2(t+1)$ and high frequency components of $\hat{z}_H(t+1|t)$ and $\hat{\sigma}_H^2(t+1)$ can be obtained via some UKFNN (Guan and et al., 2010).

4.4 Overall load forecasting and confidence interval estimation

To quantify forecasting accuracy, the confidence interval was obtained by using the neural networks trained by hybrid Kalman filters. Within the wavelet neural network framework, the covariance matrices of Kalman filters for individual frequency components contained forecasting quality information of individual load components. When load components were combined to form the overall forecast, the corresponding covariance matrices would also be appropriately combined to provide accurate confidence intervals for the overall prediction (Guan et al., 2010).

The overall load prediction is the sum of low component prediction $\hat{z}_L$, medium component prediction $\hat{z}_M$ and high component prediction $\hat{z}_H$ because these components are orthogonal based on wavelet decomposition property:

$$\hat{z}(t+1|t) = \hat{z}_L(t+1|t) + \hat{z}_M(t+1|t) + \hat{z}_H(t+1|t), \tag{14}$$

By the same token, the overall standard deviation $\hat{\sigma}(t+1|t)$ for STLF is the sum of standard deviations for low and high components:

$$\hat{\sigma}(t+1|t) = \hat{\sigma}_L(t+1|t) + \hat{\sigma}_M(t+1|t) + \hat{\sigma}_H(t+1|t), \tag{15}$$

Hence, the one sigma confidence interval for STLF can be constructed by:

$$\left[\hat{z}(t+1|t) - \hat{\sigma}(t+1|t), \hat{z}(t+1|t) + \hat{\sigma}(t+1|t)\right]. \tag{16}$$

The overall scheme of training, forecasting and confidence interval estimation is depicted and summarized in Figure 9.

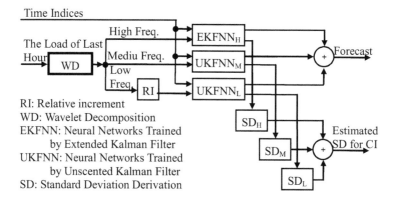

Fig. 9. Structure of a general wavelet neural networks trained by hybrid Kalman filters

4.5 Composite demand forecasting

To generate better forecasting results, a composite forecast is developed to mix multiple methods for STLF with CI estimation. The concept is based on the statistical model of ensemble forecasting to produce an optimal forecast by compositing forecasts from a number of different techniques. The method is depicted schematically in Figure 3.

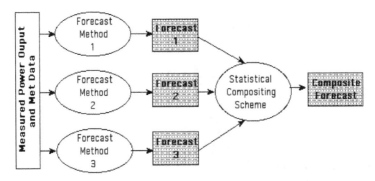

Fig. 10. Ensemble forecasting

As illustrated in Figure 11, the method runs three sample models (Forecast 1, Forecast 2 and Forecast 3) in parallel. The weights of the combination are theoretically derived based on the "interactive multiple model" approach (Bar-Shalom et al, 2001). For methods which are based on Kalman filters and have dynamic covariance matrices on the forecast load, these dynamic covariance matrices are used for the combination. Otherwise, static covariance matrices derived from historic forecasting accuracy are used instead.

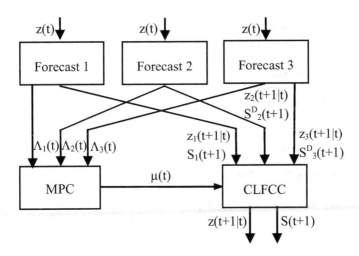

CLFCC: Composite load forecasting and covariance combination
MPC: Mixing probability calculation

Fig. 11. Structure of composite forecasting with confidence interval estimation

The relative increment (RI) in load is used to help capture the load features in the method since it removes a first-order trend and anchor the prediction by the latest load (Shamsollahi et al., 2001). After normalization, the RI in load of last time period $z(t)$ is denoted as the input to the NN, where time t is the time index. The mixing weight $\mu(t)$ can be calculated through the likelihood functions $\Lambda_j(t)$, with superscript j=1, 2, 3 representing Forecasts 1, 2 &3 respectively:

$$\Lambda_j(t) = N\left\{z(t); \hat{z}_j\left(t \mid t-1\right), S_j(t)\right]\right\}, \tag{17}$$

$$\mu_j(t) = \Lambda_j(t) \cdot \bar{c}_j \bigg/ \left(\sum_{j=1}^{3} \Lambda_j(t) \cdot \bar{c}_j \right),$$

$$where \quad \bar{c}_j = \sum_{i=1}^{3} p_{ij} \cdot \mu_i(t-1) \quad j = 1,2,3, \tag{18}$$

where p is the transition probability to be configured manually. S_1, S_2 and S_3 are sample covariance matrices from Forecasts 1, 2 &3 derived from historic forecasting accuracy. Without loss of generality, we assume that dynamic covariance matrices S_2^D for Forecast 2 and S_3^D for Forecast 3 are available. To make a stable combination, the dynamic innovation matrices S_2^D from Forecast 2 and S_3^D from Forecast 3 are not used to calculate likelihood functions Λ_2 and Λ_3 since S_2^D and S_3^D may largely affects the mixing weight. Then predictions from individual models can be combined to form the forecast:

$$\hat{z}(t+1 \mid t) = \sum_{j=1}^{3} \mu_j(t) \cdot \hat{z}_j(t+1 \mid t) \cdot \tag{19}$$

The output $z(t+1 \mid t)$ from NNs has to be transformed back due to the RI transformation on the load input. Similar to the prediction combination, the static covariance matrix S_1 derived from historic forecasting accuracy and dynamic covariance matrices S_2^D and S_3^D will also be combined. Here, S_1 S_2^D and S_3^D are the covariance matrices for NN outputs (estimated RI in load). Since RI is a nonlinear transformation, the covariance matrix has to be transformed. If S_1 S_2^D and S_3^D can be obtained directly from individual models, they can be combined first:

$$S(t+1) = \mu_1(t) \cdot S_1(t+1) + \mu_2(t) \cdot S_2^D(t+1) + \mu_3(t) \cdot S_3^D(t+1) \tag{20}$$

Then, $S(t+1)$ will be used to further derive CIs with respect to RI transformation (Guan et al., 2010).

Demand forecast and its corresponding confidence intervals are crucial inputs to the Generation Control Application which robustly dispatch the power system using a series of coordinated scheduling functions.

5. Generation control application

Generation Control Application (GCA) is an application designed to provide dispatchers in large power grid control centers with the capability to manage changes in load, generation, interchange and transmission security constraints simultaneously on a intra-day and near real-time operational basis. GCA uses least-cost security-constrained economic scheduling and dispatch algorithms with resource commitment capability to perform analysis of the desired generation dispatch. With the latest State Estimator (SE) solution as the starting point and transmission constraint data from the Energy Management System (EMS), GCA Optimization Engines (aka Scheduler or SKED) will look ahead at different time frames to forecast system conditions and alter generation patterns within those timeframes.

This rest of this section will focus on the functionality of SKED engines and its coordination with COP.

5.1 SKED optimization engine

SKED is a Mixed Integer Programming (MIP) / Linear Programming (LP) based optimization application which includes both unit commitment and unit dispatch functions. SKED can be easily configured to perform scheduling processes with different heart beats and different look-ahead time. A typical configuration for GCA includes three SKED sequences:

- SKED1 provides the system operator with intra-day incremental resource (including generators and demand side responses) commitment/de-commitment schedules based on Day-ahead unit commitment decision to manage forecasted upcoming peak and valley demands and interchange schedules while satisfying transmission security constraints and reserve capacity requirements. SKED1 is a MIP based application. It is typically configured to execute for a look-ahead window of 6-8 hours with viable interval durations, e.g., 15-minute intervals for the 1st hour and hourly intervals for the rest of study period.

- SKED2 will look 1-2 hour ahead with 15-minute intervals. SKED2 will fine-tune the commitment status of qualified fast start resources and produce dispatch contours. SKED2 also provides resource ramping envelopes for SKED3 to follow (Figure 12).

- SKED3 is a dispatch tool which calculates the financially binding base points of the next five-minute dispatch interval and advisory base-points of the next several intervals for each resource (5 min, 10 min, 15 min, etc). SKED3 can also calculate ex-ante real-time LMPs for the financial binding interval and advisory price signals for the rest of study intervals. SKED3 is a multi-interval co-optimization LP problem. Therefore, it could pre-ramp a resource for the need of load following and real-time transmission congestion management.

Traditionally, due to the uncertainty in the demand and the lack of compliance from generators to follow instructions, RTOs have to evaluate several dispatch solutions for different demand scenarios (low (L), medium (M) and high (H)). Figure 4 depicts such practice for real-time dispatch. Except for the initial conditions (e.g. MW from State Estimator (SE)), the solutions are independent. The operators have to choose to approve one of the three load scenarios based on their human judgments on which scenario is more likely to occur.

The conventional way of dealing with the demand uncertainty is stochastic optimization (Wu et al., 2007; Verbic and Cañizares, 2006; Ruiz et al., 2009). The data requirements of stochastic optimization, makes it more appropriate to solve longer term problems, e.g. expansion and operational planning, including day-ahead security constrained unit commitment process. However, the simplicity and flexibility of the solution proposed in this chapter makes it more practical for real-time dispatch.

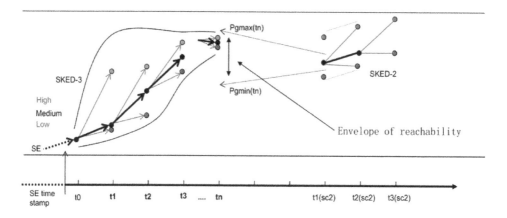

Fig. 12. SKED 2 and SKED 3 Coordination

A. Single interval dispatch model

The traditional single interval dispatch model is formulated as a Linear Programming (LP) problem:

$$Minimize$$
$$\sum_i (c_i * Pg_i)$$

$$subject\ to$$
$$\sum_i Pg_i = Dm$$
$$Pg_i^{min} \le Pg_i \le Pg_i^{max}$$
$$-(time - time_{SE}) * RRDn_i \le Pg_i - Pg_{iSE} \le (time - time_{SE}) * RRUp_{i,t}$$
$$-F_k^{max} \le \sum_i Dfax_{Fk,i} * Pg_i \le F_k^{max}$$

(21)

where

c_i Offer price for resource i

Pg_i Dispatch level for resource i

Dm Demand forecast for target time

Pg_i^{min}, Pg_i^{max} Min and max dispatch level for resource i

F_k^{max} Line/flowgate k transmission limit

$Dfax_{Fk,i}$ Sensitivity of line/flowgate k to injection i (demand distributed slack)

$time$ Target time

$time_{SE}$ State Estimator time stamp

$RRDn_i$ Maximum ramp rate down for resource i

$RRUp_i$ Maximum ram rate up for resource i

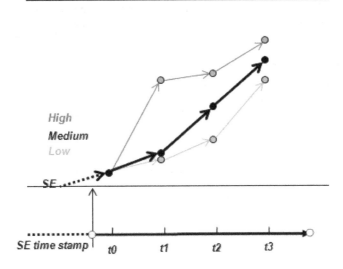

Fig. 13. Three independent dispatch solutions

B. Dynamic dispatch model

Adding the time dimension into the single interval dispatch problem above described, the basic multi-interval dispatch (dynamic dispatch) model is formulated as an extended LP problem (sub-index t is added to describe interval t related parameters and variables, as appropriate):

Minimize

$$\sum_t \left\{ \sum_i \left(c_{i,t} * Pg_{i,t}\right) * \left(Time_t - Time_{t-1}\right) / 60 \right\}$$

subject to

$$\sum_i Pg_{i,t} = Dm_t$$

$$Pg_{i,t}^{min} \le Pg_{i,t} \le Pg_{i,t}^{max} \tag{22}$$

$$-\left(time_t - time_{t-1}\right) * RRDn_{i,t} \le Pg_{i,t} - Pg_{i,t-1} \le \left(time_t - time_{t-1}\right) * RRUp_{i,t}$$

$$-F_{k,t}^{max} \le \sum_i Dfax_{Fk,i} * Pg_{i,t} \le F_{k,t}^{max}$$

$$for \quad \forall t = \in \{t1,...tn\}$$

Figure 13. illustrates the dynamic dispatch model with multiple scenario runs.

C. Robust dispatch model

A more robust solution that co-ordinates the three demand scenarios, guaranteeing the "reach-ability" of confidence interval of demand forecast from the medium demand dispatch is proposed.

The solution would provide a single robust dispatch, guaranteeing that the dispatch levels for the low and high demand scenarios can be reached from the dispatch corresponding to the medium (expected) demand scenario within consecutive intervals in the study horizon, e.g. avoiding extreme measures like demand curtailment if the high demand scenario materializes and it is too late to catch up. The robust solution proposed is depicted in Figure 14. The cost of Robust Dispatch will be higher than ordinary dispatch using medium load level. It can be justified as a type of ancillary services for load following.

A further refinement to the proposed solution is to limit the cost of the "robustness" and specify a merit order of the intervals in which robustness is more valuable.

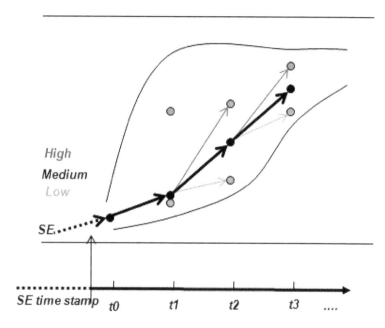

Fig. 14. Robust dispatch solution

The following LP problem co-ordinates the three demand scenarios into one "robust" solution. The objective function and constraints corresponding to the medium demand scenario are the same as those of an independent dynamic dispatch (M only); for the high and low demand scenarios, however, while the objective function terms are the same as those of independent dynamic dispatches (H only and L only), the maximum ramp rate constraints for each resource do not link dispatch levels of consecutive intervals for the same scenario (H→H, L→L); instead, for a given interval t such constraints link the H and L dispatch levels with the dispatch level corresponding to the M scenario in the preceding interval t-1 (M→H and M→L); guaranteeing the "reach-ability" of the low and high demand scenarios dispatches from the medium demand dispatch level in successive intervals (upper and lower case h, m and l are used as extensions to describe high, medium and low demand scenarios parameters and variables).

Minimize

$$\sum_t \left\{ \sum_i \left(c_{i,t} * Pgm_{i,t} \right) * (time_t - time_{t-1}) / 60 \right\}$$

$$+ \sum_t \left\{ \sum_i \left(c_{i,t} * Pgh_{i,t} \right) * (time_t - time_{t-1}) / 60 \right\}$$

$$+ \sum_t \left\{ \sum_i \left(c_{i,t} * Pgl_{i,t} \right) * (time_t - time_{t-1}) / 60 \right\}$$

subject to

$$\sum_i Pgm_{i,t} = Dm_t$$

$$Pg_{i,t}^{min} \leq Pgm_{i,t} \leq Pg_{i,t}^{max}$$

$$-(time_t - time_{t-1}) * RRDn_{i,t} \leq Pgm_{i,t} - Pgm_{i,t-1} \leq (time_t - time_{t-1}) * RRUp_{i,t}$$

$$-F_{k,t}^{max} \leq \sum_i Dfax_{Fk,i,t} * Pgm_{i,t} \leq F_{k,t}^{max}$$

$$\sum_i Pgh_{i,t} = Dh_t$$

$$Pg_{i,t}^{min} \leq Pgh_{i,t} \leq Pg_{i,t}^{max}$$

$$-(time_t - time_{t-1}) * RRDn_{i,t} \leq Pgh_{i,t} - Pgm_{i,t-1} \leq (time_t - time_{t-1}) * RRUp_{i,t}$$ (23)

$$-F_{k,t}^{max} \leq \sum_i Dfax_{Fk,i,t} * Pgh_{i,t} \leq F_{k,t}^{max}$$

$$\sum_i Pgl_{i,t} = Dl_t$$

$$Pg_{i,t}^{min} \leq Pgl_{i,t} \leq Pg_{i,t}^{max}$$

$$-(time_t - time_{t-1}) * RRDn_{i,t} \leq Pgl_{i,t} - Pgm_{i,t-1} \leq (time_t - time_{t-1}) * RRUp_{i,t}$$

$$-F_{k,t}^{max} \leq \sum_i Dfax_{Fk,i,t} * Pgl_{i,t} \leq F_{k,t}^{max}$$

for $\forall t = \in \{t1,...tn\}$

It is important to note that there is certainly a tradeoff between cost and robustness for any given robust dispatch solution using the methodology proposed above. Figure 15 illustrates the conceptual idea of relationship between cost and flexibility which is proportional to robustness. The value of the "Δcost acceptable" will be very much dependent on the amount of risk one is willing to take for reliability purposes when dispatching the system.

5.2 SKED and COP coordination

GCA is built upon a modular and flexible system architecture. Although different SKED processes are correlated, they do not replay on each other. The orchestration between SKEDi is managed by COP. This design enables low-risk, cost-effective business process evolution.

It also ensures high availability for the mission critical real-time GCA SKED functions. Failure of any one or more SKED components will cause smooth degradation of, instead of abrupt service interruptions to, real-time dispatch instructions.

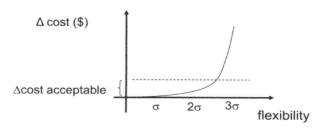

Fig. 15. Robustness vs. Cost

COP is the repository of all operating plans in a multi-stage decision process. Each SKEDi in the decision process generates a set of schedules that are reflected in its corresponding COP (COPi). The aggregated results from the multi-stage decision process are captured in the total COP (COPt), which is the consolidated outcome of the individual COPi's. SKED and COP coordination is illustrated in Figure 16.

Initialization of the COP for each operating day begins with the day-ahead schedule, which is based on the DAM financial schedules and then updated with Reliability Commitment results. Before any SKEDi is run in the current day of operation, the overall COPt is initialized with the day-ahead schedules. When COPt is suitably initialized, it will be used to generate input data for SKED1, SKED2 and SKED3. Results of SKEDi's are then used to update their respective subordinate COPi, which will cause COPt to be updated, and thus the overall iterative process continues.

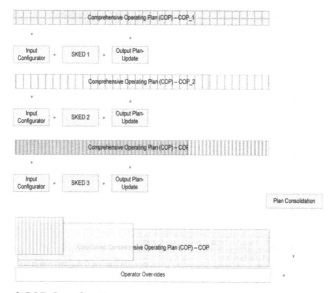

Fig. 16. SKED and COP Coordination

GCA aims at enhancing operators' forward-looking view under changing system conditions (generation capacity, ramp capability, transmission constraints, etc.) and providing operators with a "radar-type" of recommendation of actions such as startup or shutdown of fast-start resources in near real-time. As shown in Figure 17 – COP review display, various startup and shutdown recommendations are approaching the "now" timeline like an 1-dimensional radar sorted by likelihood ranking from top to bottom. The COP review display also shows actual system total generation and comparing against demand forecast and system ramp constrained capacity. This provides situation awareness of any potential abrupt ramping events or potential system imbalance and alerts operators in advance if any actions need to be taken.

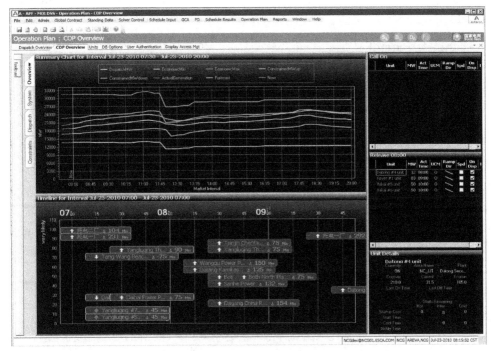

Fig. 17. Forward-looking view presented by COP Overview

6. Conclusion

Significant capacity of renewable generation resources operating online at any given time is of great concern to grid security due to the intermittent nature of many of the resources. On one hand, the potential volatility of the intermittent generation output could cause great stress on the system's generation planning and ramp management. On the other hand, these intermittent resources could be operating at locations that contribute to transmission line congestion and become very challenging problems for a lot of the RTOs.

This chapter addresses the challenges of renewable integration from a generation dispatch perspective. The framework of Smart Dispatch is proposed in which the applications of demand forecasting and robust dispatch are discussed in detail. A new dispatch system called Generation Control Application (GCA) is described to address the challenges posed by renewable energy integration. GCA aims at enhancing operators' forward-looking view under changing system conditions such as wind speed or other weather conditions. GCA provides operators with situation awareness of any potential abrupt ramping events or potential system imbalance and alerts operators in advance if any actions need to be taken. With dynamic and robust dispatch algorithm and flexible system configuration, the system provides adequate system ramping capability to cope with uncertain intermittent resources while maintaining system reliability in large grid operations.

Smart Dispatch is deemed critical for the success of efficient power system operations in the near future.

7. Acknowledgment

The views expressed in this chapter are solely those of the author, and do not necessarily represent those of Alstom Grid.

8. References

Amjady, N.; Keynia, F. &Zareipour, H. (2010). Short-Term Load Forecast of Microgrids by a New Bilevel Prediction Strategy, *Smart Grid, IEEE Transactions on*, vol.1, no.3, pp.286-294, Dec. 2010.

Bar-Shalom, Y.; Li, X. R. & Kirubarajan, T. (2001). Estimation with Applications to Tracking and Navigation: Algorithms and Software for Information Extraction, *J. Wiley and Sons*, pp.200-209 and pp.385-386.

Chen, Y.; Luh, P. B.; Guan, C.; Zhao, Y. G.; Michel, L. D.; Coolbeth, M. A.; Friedland, P. B. & Rourke, S. J. (2010). Short-term Load Forecasting: Similar Day-based Wavelet Neural Networks, *IEEE Transaction on Power Systems*, 25(1):322-330.

Cheung, K. W.; Wang, X.; Chiu, B. -C.; Xiao, Y. & Rios-Zalapa, R. (2010). Generation Dispatch in a Smart Grid Environment, *Proceedings of 2010 IEEE/PES Innovative Smart Grid Technologies Conference (ISGT 2010)*.

Cheung, K. W.; Wang, X. & Sun, D. (2009). Smart Dispatch of Generation Resources for Restructured Power Systems, *Proceedings of the 8th IET International Conference on Advances in Power System Control, Operation and Management*.

Chow, J. H.; deMello, R. & Cheung, K. W. (2005). Electricity Market Design: An Integrated Approach to Reliability Assurance. (Invited paper), *IEEE Proceeding (Special Issue on Power Technology & Policy: Forty Years after the 1965 Blackout)*, vol. 93, pp.1956-1969, Nov. 2005.

Chuang, A. S. & Schwaegerl, C. (2009). Ancillary services for renewable integration, *Integration of Wide-Scale Renewable Resources into the Power Delivery System, 2009 CIGRE/IEEE PES Joint Symposium*.

DOE's Report (2010). Incorporating Wind Generation and Load Forecast Uncertainties into Power Grid Operations, PNNL-19189.

Gisin, B.; Qun Gu; Mitsche, J.V.; Tam, S. & Chen, H. (2010). "Perfect Dispatch" – as the Measure of PJM Real Time Grid Operational Performance, *Proceedings of the IEEE PES 2010 General Meeting*.

Guan, C.; Luh, P. B.; Cao, W.; Zhao, Y.; Michel, L. D. & Cheung, K. W. (2011). Dual-tree M-band Wavelet Transform and Composite Very Short-term Load Forecasting, *Proceedings of the IEEE PES 2011 General Meeting.*

Guan, C.; Luh, P. B.; Coolbeth, M. A.; Zhao, Y.; Michel, L. D.; Chen, Y.; Manville, C. J.; Friedland, P. B. & Rourke, S. J. (2009). Very Short-term Load Forecasting: Multilevel Wavelet Neural Networks with Data Pre-filtering, *Proceedings of the IEEE PES 2009 General Meeting.*

Guan, C.; Luh, P.B.; Michel, L.D.; Coolbeth, M.A. & Friedland, P.B. (2010). Hybrid Kalman algorithms for very short-term load forecasting and confidence interval estimation, *Power and Energy Society General Meeting, 2010 IEEE* , vol., no., pp.1-8, 25-29 July 2010.

Julier, S. J.; Uhlman, J. K. & Durrant-Whyte, H. F. (1995). A New Approach for Filtering Nonlinear Systems, *In Proc. American Control Conf, Seattle, WA*, 1628-1632.

Ma, X.-W.; Sun, D. & Cheung, K. W. (1999). Energy and Reserve Dispatch in a Multi-Zone Electricity Market, *IEEE Transaction of Power Systems*, vol. 14, pp.913-919, Aug. 1999.

Meyer, P. L. (1970). *Introductory Probability and Statistical Applications*, Addison-Wesley Publishing Company, Inc., 1970.

Puskrius, G. V. & Feldkamp, L. A. (1991). Decoupled Extended Kalman Filter Training of Feedforward Layered Networks, *Proceedings of IEEE International Joint Conference on Neural Networks*, 771-777.

Rios-Zalapa, R.; Wang, X.; Wan, J. & Cheung, K. W. (2010). Robust Dispatch to Manage Uncertainty in Real Time Electricity Markets, *Proceedings of 2010 IEEE/PES Innovative Smart Grid Technologies Conference (ISGT 2010).*

Ruiz, P. A.; Philbrick, C. R.; Zak, E.; Cheung, K. W. & Sauer, P. W. (2009). Uncertainty Management in the Unit Commitment Problem, *IEEE Transactions on Power Systems*, Vol. 24, No. 2, May 2009, pp. 642-651.

Schweppe, F. C.; Caramanis, M. C.; Tabors, R. D. & Robn, R. E. (1998). *Spot Pricing of Electricity*, Kluwer Academic Publishers, 1998.

Shamsollahi, P.; Cheung, K. W.; Chen, Q. & Germain, E. H. (2001). A Neural Network Based Very Short term Load Forecaster for the Interim ISO New England Electricity Market System, *Proceedings of the 22nd International Conference on Power Industry Computer Applications*, pp.217-222 (2001).

Singhal, S. & Wu, L. (1989). Training feed-forward networks with the extended Kalman filter, *Proceedings of IEEE International Conference on Acoustics Speech and Signal Processing*, 1187–1190.

Verbic, G. & Cañizares, C. A. (2006). Probabilistic Optimal Power Flow in Electricity Markets Based on a Two-Point Estimate Method, *IEEE Transactions on Power Systems*, Vol. 21, No. 4, November 2006, pp. 1883-1893.

Wood, A. J. & Wollenberg, B. F. (1996). *Power Generation, Operation, and Control*, second edition, John Wiley & Sons, New York, 1996.

Wu, L.; Shahidehpour, M.; Li, T. (2007). Stochastic Security-Constrained Unit Commitment, *IEEE Transactions on Power Systems*, Volume: 22, No. 2, May 2007.

Zhang, H.-T.; Xu, F.-Y. & Zhou, L. (2010). Artificial neural network for load forecasting in smart grid, *Machine Learning and Cybernetics (ICMLC), 2010 International Conference on*, vol.6, no., pp.3200-3205, July 11-14, 2010.

Zhang, L. & Luh, P. B. (2005). Neural Network-based Market Clearing Price Prediction and Confidence Interval Estimation with an Improved Extended Kalman Filter Method, *IEEE Transactions on Power Systems,* Vol. 20, No. 1, February, 2005, pp. 59-66.

5

Renewable Energy from Palm Oil Empty Fruit Bunch

Somrat Kerdsuwan and Krongkaew Laohalidanond
The Waste Incineration Research Center,
Department of Mechanical and Aerospace Engineering,
King Mongkut's University of Technology North Bangkok
Thailand

1. Introduction

The world economy is the key driver for global primary energy consumption. According to the economic recovery in 2010, the world primary energy consumption is rebounded which was accounted for 12,002.4 Mtoe, increased by 28 % from year 1998 (BP, 2010). The important primary energy sources are fossil fuels, e.g. crude oil, natural gas, coal, whereas the renewable energy source has a small share. If the world population, world economy and consequently the world energy consumption are still growing rapidly, in spite of there are more than 1,000 thousands million barrel proved oil reserved in 2010 (BP, 2010), the world will face to problem of oil price crisis and also oil diminishment in the near future. Beyond that crisis, the utilization of fossil fuel will lead to an increase in carbon dioxide emission in the atmosphere. In 2007, global carbon dioxide emission was accounted for 29.7 billion metric ton (U.S. Energy Information Administration [US IEA], 2010), increased approximately 1.7 % from previous year (29.2 billion metric ton in 2006 (US IEA, 2006)) and it was predicted that the concentration of carbon dioxide still increases by an average of 0.1 % annually for OECD countries and an average of 2.0 % per year for non-OECD countries until 2035 (US IEA, 2010). The high concentration of carbon dioxide causes the greenhouse effect and consequently, the global warming which becomes the serious problem to human race.

Nowadays there are many efforts to reduce both the use of fossil fuel and the carbon dioxide emission by using renewable energy as a substitute to fossil fuel. This renewable energy includes solar energy, wind energy, hydro energy as well as energy derived from biomass, tide and geothermal. These renewable energy sources can be used not only for heat and power generation but also for liquid transportation fuel production. During four consecutive years from end-2004 to end-2008, the global solar photovoltaic had increased six-fold to more than 16 GW_e, while the global wind power generation had risen by 250 % to 121 GW_e. More than 280 GW_e had been produced from hydro, geothermal and biomass power plants which increased 75 % from the last four year. In the same time interval, the heat production from solar heating was also doubled to 145 GW_{th}. In the perspective of biofuel production, biodiesel production increased six-fold to 12 billion liters per year, whereas ethanol production was doubled to 67 billion liters per year (Renewable Energy Policy Network for the 21st Century [REN21], 2009).

Analogy to global energy situation, Thailand has faced to the problem of energy crisis, especially oil price crisis. Thailand has been relying on fossil fuel as the primary source of energy which has to be imported from foreign countries, and simultaneously Thailand energy consumption has been increased rapidly and continuously. This chapter will focus on the energy situation in Thailand, the energy policy plan and finally the possibility of using renewable energy sources as alternative energy to energy from fossil fuel.

1.1 Energy situation in Thailand

Thailand is a developing country located at the middle of Southeast Asia with a population of 63.5 million in 2009 (Department of Provincial Administration [DOPA], 2010), increased 2.3 % from 2005. During the same period, the Gross Domestic Product (GDP) increased from 7,092,893 million Baht in 2005 to 9,041,551 million Baht in 2009 (Office of National Economic and Social Development Board [ONESDB], 2009) (1 US$ ~ 31 Baht) or equal to the increasing rate of 2.7 %. There exists a two way casual relationship between economic/population growth and energy consumption in case of Southeast Asia (Wianwiwat & Adjaye, 2011). That is the higher level of economic growth and population growth will result in the higher energy consumption. Identical to other countries in Southeast Asia, the economic and industrial developments as well as the growth in population lead to the higher energy consumption. In 2009, domestic final energy consumption was accounted for 69,177 ktoe (Energy Policy and Planning Office [EPPO], 2010) which increased 2.9 % from previous year. Table 1 shows the final energy consumption in Thailand from 2005 to 2009.

Year	Consumption (ktoe)	Increasing Rate (%)
2005	63,061	-
2006	62,904	-0.25
2007	65,950	4.62
2008	67,256	1.98
2009	69,177	2.86

Table 1. Domestic final energy consumption from 2005 to 2009 in Thailand (EPPO, 2010)

The primary energy sources consumed in Thailand are mostly derived from fossil fuels, e.g. petroleum, natural gas and coal which contributed to 56,693 ktoe in 2009 (EPPO, 2010) or equivalent to 82 % of the final energy consumption, whereas renewable energy covered only 18 %. Figure 1 illustrates the final energy consumption by types in 2009.

Most of commercial energy source used in Thailand is petroleum product which was equal to 31,959 ktoe in 2009. Diesel, gasoline and liquid petroleum gas are the major petroleum products used in transportation sector. Diesel and gasoline consumption accounted for the share of 46 % and 19 %, respectively, whereas liquid petroleum gas consumption was amounted to 17 % (EPPO, 2010). Natural gas was mainly used as fuel for power generation, holding a share of 68 % of the total consumption in 2009. The remainder was used in gas separation plant, with the share of 17 %; in industries with 11 % and in transportation sector for 4 % (EPPO, 2010). The share of coal and lignite consumption as fuel in power generation sector was almost at the same level as its consumption in industrial sector.

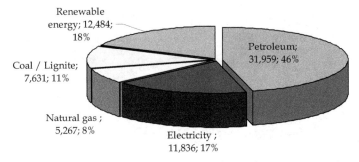

Fig. 1. Final energy consumption in 2009 by types in ktoe; percent (EPPO, 2010)

Unfortunately, Thailand has insufficient crude oil or good quality fossil fuels, these energy sources have to be imported from foreign countries, especially Middle East countries for oil import. In 2009, Thailand imported a substantial amount of commercial energy sources, approximately 59,333 ktoe which costed more than 760 billion Baht (EPPO, 2010). Figure 2 shows the expense for final energy import in 2009.

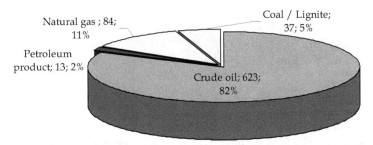

Fig. 2. Final energy consumption in 2009 by types in billion Baht; percent (EPPO, 2010)

Considering greenhouse gas (GHG) emission, carbon dioxide emission in Thailand during the last five years increased from 192,486 kton in 2005 to 208,476 kton in 2009 which is listed in more detail in Table 2.

Year	Amount (kton)	Increasing Rate (%)
2005	192,486	-
2006	193,136	0.34
2007	200,439	3.78
2008	203,181	1.37
2009	208,476	2.61

Table 2. Carbon dioxide emission from 2005 to 2009 in Thailand (EPPO, 2010)

The main sources of carbon dioxide emission are oil, natural gas and coal which are significantly used as fuel for energy production via combustion process. It is expressed that the direct combustion of these fossil fuel is the largest source of GHG emission from human activities (Sawangphol & Pharino, 2011). These fossil fuels are used in industrial, power generation and transportation sector. Figure 3 and Figure 4 show the amount of carbon dioxide emission in 2009 by sources and by sectors, respectively.

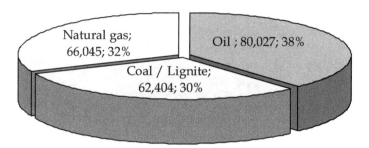

Fig. 3. Carbon dioxide emission in 2009 in Thailand by sources in kton (EPPO, 2010)

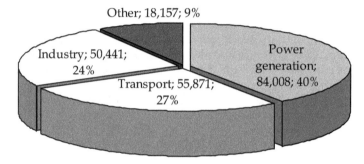

Fig. 4. Carbon dioxide emission in 2009 in Thailand by sectors, in kton (EPPO, 2010)

1.2 Energy policy and plan

Since Thailand has spent a large amount of money for importing commercial energy sources, Thai government as well as private and public organizations have realized about that, Ministry of Energy of Thailand has promoted the use of renewable energy, including biomass, municipal solid waste, biogas, wind and solar power for power generation or transportation fuel production by announcing the 15-Years of Alternatives Energy Development Plan (AEDP, 2009) on January 28, 2009. The objective of this AEDP is to strengthen and promote the utilization of renewable energy in order to replace the oil import. The main target of AEDP is to increase the portion of using alternative energy to 20 % of national final energy consumption by 2020. The plan will be implemented into three phases: short-term from 2008 to 2011, mid-term from 2012 to 2016 and long-term from 2017 to 2020.

From Table 3, it can be noticed that the main target of AEDP until 2020 is the utilization of biomass for electricity and heat production. In addition to electricity and heat, biomass can also be used as feedstock for biofuel production. Considering cost of electricity and heat generated from renewable energy, electricity and heat produced from solar and wind energy has a higher cost with more complicated technology for Thailand than electricity and heat from produced from biomass (Bull, 2001; Owen, 2006).

Besides the reduction of primary energy sources importing, using biomass as alternative fuel can contribute to the reduction of GHG emission, since biomass is carbon neutral which emits no net carbon dioxide. Therefore, this study will emphasize only on the energy production from biomass.

	Potential	Existing	2008-2011		2012-2016		2017-2020	
Electricity	MW	MW	MW	ktoe	MW	ktoe	MW	ktoe
Solar	50,000	32	55	6	95	11	500	56
Wind	1,600	1	150	17	400	45	700	78
Hydro	700	50	165	43	281	73	324	85
Biomass	4,400	1,597	2,800	1,463	3,235	1,682	3,700	1,933
Biogas	190	29	60	27	90	40	120	54
MSW	320	5	100	60	130	87	160	96
Hydrogen			0	0	0	0	3.5	1
Total		1,714	3,330	1,616	4,231	1,938	5,508	2,303
Heat	ktoe	ktoe		ktoe		ktoe		ktoe
Solar	154	2.3		5		17		34
Biomass	7,400	2,344		3,544		4,915		6,725
Biogas	600	79		470		540		600
MSW	78	1		16		25		35
Total		2,426.3		4,035		5,497		7,394
Biofuels	Ml/day	Ml/day	Ml/day	ktoe	Ml/day	ktoe	Ml/day	ktoe
Ethanol	3.30	1.00	3	816	6.20	1,686	9	2,447
Biodiesel	3.30	1.39	3	944	3.64	1,145	4.50	1,416
Hydrogen			0	0	0	0	0.1 M kg	124
Total			6	1,760	9.84	2,831	13.50	3,987
Total energy demand (ktoe)	65,420			72,539		88,389		112,046
Total renewable energy demand (ktoe)	3,411.80			7,411		10,266		13,684
Portion of renewable energy use	5.2 %			10.2 %		11.6 %		12.2 %
Natural gas (mmscfd)	91.5	345	3,045	826	7,290	1,035	9,135	
Total alternative energy demand (ktoe)				10,456		17,556		22,819
Portion of alternative energy used				14.4 %		19.9 %		20.4 %

Table 3. 15-Years of Alternatives Energy Development Plan (AEDP, 2009)

1.3 Biomass potential in Thailand

Since Thailand is the agricultural base country, there are a lot of agricultural crops, e.g. paddy rice, sugarcane, cassava and palm oil. During the harvesting and processing of these agricultural crops, some residues are left over, e.g. rice straw and rice husk from paddy rice, bagasse and sugarcane leave from sugarcane, cassava rhizome from cassava as well as palm oil shell, palm oil fiber and palm oil empty fruit bunch from palm oil fruit. These residues can further be used as the substitute to fossil fuel for energy production and, consequently, can solve the problem of high energy price as well as global warming. The amount of residues from these agricultural products can be estimated by their productivities, Crop-to-Residual-Ratio (CRR) and Surplus-Availability-Factor (SAF). The CRR is expressed as the amount of residues generated per 1 unit mass of an agricultural product and the SAF is the amount of unused residues or residues left-over which are not used for any purposes. The potential of bio-energy from these agricultural products is then calculated from the quantity of biomass residues and the lower heating value of biomass. The office of Agricultural Economic reported the production of four main agricultural products in 2009 as followed: paddy rice 32,116 kton, sugarcane 68,808 kton, cassava 22,006 kton and palm oil fruit 8,223 kton (OAE, 2010). Table 4 shows the amount of residues and energy potential from domestic main agricultural products based on productivity in 2009.

Agricultural product	Residues	Productivity (kton)	CRR (Papong et al., 2004)	SAF (Papong et al., 2004)	Quantity of residues (kton)	LHV (MJ/kg) (Prasertsan & Sajjakulnu kit, 2006)	Total Energy (PJ)
Paddy rice		32,116					
	Rice husk		0.23	0.493	3,641.63	14.27	51.97
	Rice straw		0.447	0.684	9,819.40	10.24	100.55
Sugarcane		68,808					
	Bagasse		0.291	0.227	4,545.25	8.31	37.77
	Sugarcane leaves		0.302	0.986	20,489.10	8.70	178.26
Cassava		22,006					
	Cassava rhizome		0.49	0.98	10,567.28	5.50	58.12
Palm oil fruit		8,223					
	Shell		0.049	0.037	14.91	18.46	0.28
	Fiber		0.147	0.134	161.98	17.62	2.85
	Empty fruit bunch		0.250	0.584	1,200.56	17.86	21.44
	Frond		2.604	1.00	21,412.69	9.83	210.49
Total							661.73

Table 4. Energy potential of main agricultural residues in Thailand

From Table 4, the energy potential of agricultural residues generated from four main agricultural products in 2009 was accounted for 661.73 PJ. The energetic potential of rice straw, sugarcane leaves and palm oil frond is very high compared to other types of biomass, 100.55 PJ, 178.26 PJ and 210.49 PJ, respectively; however the utilization of these residues as a renewable source for energy production has hardly been found. Rice straw is normally used for animal fodder, soil cover material and paper industry, while palm oil frond is also served as soil cover material. Sugarcane leave is normally left in field and burnt. Currently only rice husk, bagasse and palm oil shell are widely used as feedstock in stand-alone or co-firing power plant for heat and power production (Papong et al., 2004; Prasertsan & Sajjakulnukit, 2006). Not only for heat and power production, rice husk, bagasse and palm oil shell can be used in several industries, e.g. animal fodder and paper industry. Despite there are a large quantity of sugarcane leave and cassava rhizome, their heating value is very low compared to other biomass. This study will not focus on such biomass. Although the quantity of Palm Oil Empty Fruit Bunch (PEFB) available is in the sixth rank of all biomass, the energetic quality is high due to its high heating value and it is not yet mainly used as alternative fuel due to its high moisture and volatile matter with low ash melting temperature, therefore the scope of this study focuses on renewable energy utilization from PEFB.

2. Utilization of palm oil empty fruit bunch

The main content of this following section deals with the utilization of palm oil empty fruit bunch as renewable energy. However, this section will provide an important information about palm oil and its plantation as well as a brief discussion about its utilization for non-energetic purposes prior to go deeply in more detail of its utilization for energetic purposes.

2.1 Oil palm plantation

There are two families of oil palm, *Elaeis guineensis* which is native to western Africa and *Elaeis oleifera* whose origin is in tropical Central America and South America. The palm family which is widely cultivated in Thailand is Elaeis guineensis. It was first introduced to Thailand in 1968 (Prasertsan & Sajjakulnukit, 2006). Nowadays, the plantation of palm oil in Thailand is continuously increased because Thai government announced the policy of producing palm oil based biodiesel as renewable energy, as already mention in Table 3. The Office of Agricultural Economics (OAE, 2010) reported the oil palm plantation area in 2009 was accounted for 3,165,000 Rai (1 Rai = 1,600 m²) which increased by 56 % from the last five years and the target of 10 million Rai should be achieved by 2029 (Yangdee, n.d.). The oil palm production is increased by 64 % from 5,003,000 ton in 2005 to 8,223,000 ton in 2009. More than 90 % of palm oil plantation area in Thailand is located in Southern part of Thailand, especially in Chumporn, Surat Thani and Krabi.

Elaeis guineensis is vertical trunk and the feathery nature of leaves. There are 20-40 new leaves, called "frond" developed each year. The fruit bunches develop between trunk and base of the new fronds. Typically the first commercial crop can be harvested after 5-6 years of plantation and can provide fresh fruits for 25-30 years (Perez, 1997). The weight of compact fruit can varies from 10 to 40 kilograms. Each fruit is sphere in shape, dark purple, almost black before it ripens and turns to orange-red when ripe (Katamanee, 2006). Figure 5 illustrates oil palm tree and fresh fruit bunch.

2.2 Palm oil empty fruit bunch (PEFB)

Palm oil empty fruit bunch (PEFB) is waste residue generated from palm oil industries. After harvesting fresh fruit bunches from oil palm tree, these bunches are sterilized in a horizontal steam sterilizer to inactivate enzymes present in pericarp and loosen fruits from bunches. The sterilized bunches are fed into a rotary drum thresher in order to remove the sterilized fruit from bunches. These bunches without fruit are called as empty fruit bunch (EFB) which are conveyed to the damping ground, whereas the sterilized fruits are further used as feedstock for palm oil production in palm oil extraction process by means of screw type press. The effluents from screw type press are nuts and fibers which are separated from each other by cyclone. After this separation, nuts are cracked into shells and kernels. The former are solid waste and left unused, the latter are sent to the kernel oil mill (Mahlia et al., 2001; Prasertsan & Sajjakulnukit, 2006). It was reported that 20-22 tons of empty fruit bunch, 14 tons oil-rich fiber and 5 tons of shell are generated from 100 ton of fresh fruit bunch (Perez, 1997; Katamanee, 2006), as illustrated in Figure 6.

Source: Available from http://gardendoctor.files.wordpress.com/2010/05/elaeis_guineensis.jpg (1 Sep 2010)

Fig. 5. Oil Palm Tree and fresh fruit bunch

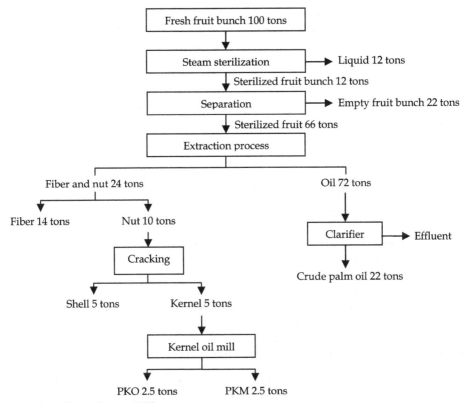

Source: Adapted from (Perez, 1997)

Fig. 6. Palm Oil Empty Fruit Bunch as waste from Palm mill industry

PEFB is a dry brown bunch with non-uniform shape and low bulk density. Its length and width depend on the size of fresh fruit bunch and can vary from 17-30 cm long and 25-35 cm wide.

Fig. 7. Palm Oil Empty Fruit Bunches

Elemental compositions and some properties of PEFB are different by sources of feedstock. Table 5 shows and compares proximate and ultimate analysis of PEFB from Thailand and Malaysia.

		Thailand (air-dried) (Own investigation)	Malaysia (air-dried) (Hamzah, 2008)
Proximate analysis			
Moisture	% wt.	8.34	8.75
Volatile matter	% wt.	73.16	79.65
Fix carbon	% wt.	12.20	8.60
Ash	% wt.	6.30	3.00
Ultimate analysis			
C	% wt.	43.8	48.79
H	% wt.	6.20	7.33
O	% wt.	42.64	40.18
N	% wt.	0.44	0
S	% wt.	0.09	0.68
Others	% wt.	0.53	0.02
Ash	% wt.	6.30	3.00
Lower Heating Value	MJ/kg	19.24	18.96

Table 5. Proximate and ultimate analysis of PEFB from different countries

From Table 5, it can be noticed that there is no significant difference in composition of PEFB by sources. Moisture content is measured to be approximately 8 % wt. based on air dried basis. Volatile matter and fix carbon varied from 73 to 80 % wt. and 8 to 12 % wt., respectively, while ash content in PEFB from Thailand is higher than ash content in PEFB from Malaysia. The chemical composition of PEFB from Thailand calculated by ultimate analysis is $C_{3.6}H_{6.2}O_{1.3}$, whereas the chemical composition of PEFB from Malaysia is $C_{4.1}H_{7.3}O_{1.3}$. The lower heating value of PEFB from sources is almost identical and is around 19 MJ/kg.

2.3 Utilization of PEFB
The utilization of PEFB can be divided into two groups: PEFB utilization for non-energetic purposes and PEFB utilization for energetic purposes.

2.3.1 PEFB utilization for non-energetic purposes
PEFB contains a variety of nutrients, e.g. phosphorus (P), potassium (K), magnesium (Mg), Nitrogen (N), etc. It is reported that the nutrients in PEFB consist of 0.06 % P, 2.4 % K, 0.2 % Mg and 0.54 % N (Heriansyah, n.d.; Prasertsan & Sajjakulnukit, 2006). As a result of this, PEFB is a good source of organic matter. By which PEFB is widely used in Thailand as a substrate for mushroom cultivation and as an organic mulch as well as supplementary fertilizer for oil palm plantation. As a substrate for mushroom cultivation, PEFB is pressed in a rectangular block and mushroom spores are inoculated into PEFB block. Finally the block is covered by plastic sheet to maintain moisture content and limit sunlight. PEFB

mulching material on soil surface for oil palm plantation can reduce soil temperature and conserve soil moisture to improve growth and crop yield. The residue from mushroom cultivation or mulching material for oil palm plantation is the composting PEFB which can further be served as organic fertilizer. After a long duration of composting, the nutrients containing in PEFB will substantially increase, e.g. after 32 weeks composting N, P and Mg in PEFB increase from 0.54 %, 0.06 % and 0.19 % to 2.22 %, 0.355 % and 0.67 %, respectively (Heriansyah, n.d.). Another possibility is the utilization of PEFB ash as fertilizer or soil conditioner. However, this method is non-preferable because white smoke caused from high moisture content in PEFB has an aesthetic effect to the environment (Yusoff, S., 2006).

Since the PEFB has a highest fiber yield and its fibers are clean, biodegradable and compatible than other wood fibers, besides using for agricultural purposes, fiber of PEFB can be served as raw material for pulp and paper, fiberboard, mattress, cushion, building material, etc. (Law et al., 2007; Nasrin et al., 2008; Prasertsan & Sajjakulnukit, 2006; Ramli et al., 2002). Anyways, there are some limitations of using PEFB fiber. The fiber must be dried to the moisture content of 15 % and oil content has to be removed from fiber in order to improve the mechanical and physical properties of PEFB fiber.

Although, there are many applications of PEFB fiber, the utilization of PEFB fiber as raw material receives only few interests in Thailand, because there are a lot of agricultural products used for this purpose available and gain the technical knowledge in commercial scale, such as eucalyptus for pulp and paper industry, rubber wood or other wood species for fiberboard and particle board, coconut fiber for mattress.

2.3.2 PEFB utilization for energetic purposes

According to the increase in crude oil price, depletion of crude oil reserves and environmental concerns, especially global warming, many studies focus on the attempt of looking for alternative energy sources to partly replace crude oil. There are many processes for converting biomass to energy, including mechanical process, thermo-chemical process and biological process, as shown in Figure 8.

In mechanical process, as received PEFB is dried, grounded and fed into pelletting/briquetting machine in order to produce pellets/briquettes. These PEFB pellets/briquettes has good properties for using as fuel in conventional stove or co-firing plant compared to as received PEFB. It exhibits high energy content due to the decrease in moisture content, uniform size and superior combustion behavior as well as high mechanical strength (Nasrin et al., 2008).

Although there are several pathways of thermo-chemical process, the direct use of PEFB in commercial plants has rarely been found. Currently, thermo-chemical technology for converting PEFB into energy has been studied in laboratory or bench scale. Many studies nowadays focus on the pyrolysis of PEFB for bio-oil production in lab-scale, both fixed bed and fluidized bed reactor and it can be evidenced that the maximum yield of bio-oil produced from pyrolysis of PEFB without catalysts occurs at the temperature of about 500 °C and the lower heating value of bio-oil is approximately 20 MJ/kg (Abdullah & Bridwater, 2006; Abdullah et al., 2007; Azizan et al., 2009; Sukiran et al., 2009; Yang et al., 2006). The utilization of catalyst can promote the pyrolysis reaction and the maximum bio-oil can be obtained at lower temperature of about 300-350 °C with shorter residence time (Amin & Asmadi, 2008). Due to its higher moisture and ash content, lower energy content compared to palm oil shell or some types of biomass feedstock as well as its non-uniform shape (Knoef, 2005), there are a large amount of experimental studies and modeling focusing on combustion, co-firing as well as gasification of PEFB in order to investigate the feasibility of using above mentioned technologies to convert PEFB into energy in term of operating conditions, configuration of reactor, emission and efficiency (Hussain et al., 2006).

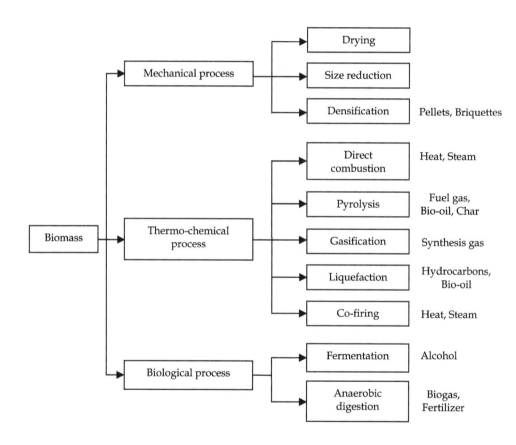

Fig. 8. Pathway for energetic utilization of biomass

3. Experimental study of PEFB gasification in a lab-scale downdraft gasifier

As mentioned in section 1.3, there is a large amount of PEFB leftover at the palm oil mills. Gasification is the prominent technology to recover energy from PEFB for heat and power production. Gasification is a thermo-chemical process to convert solid fuel into combustible gases under the sub-stoichiometric condition. The gasifier can be classified into 3 main types by the moving characteristic of bed material: fixed bed, fluidized bed and entrained flow gasifier. Table 6 shows the operating conditions of each gasifier type.

Parameter	Unit	Fixed bed	Fluidized bed	Entrained flow
Temperature	°C	300-900	700-900	1,200-1,600
Pressure	MPa	0-5	0.1-3	0.1-3
Gasification agent	-	Air/steam-oxygen mixture	Air/steam-oxygen mixture	Oxygen
Reaction time	s	600-6,000	10-100	≤ 0.5
Particle size	mm	1-100	1-10	≤ 0.5

Table 6. Operating condition for each gasifier type (Schaidhauf, 1998).

In this study, a fixed bed downdraft gasifier was chosen as the suitable technology because it has the following advantages (Belgiorno et al., 2003; Knoef, 2005; Laohalidanond, 2008; Obernberger & Thek, 2008):
- Variable fuel size (1-100 mm) and wide range of moisture content in fuel (< 20 %),
- Clean producer gas with the lowest tar content among other types of gasifier,
- Suitability for 1 MW electricity,
- Ease for operation and
- Low investment and operating cost

The experiments about PEFB gasification were carried out in a 10 kg/hr downdraft gasifier in order to identify the behavior of producer gas formation in term of producer gas yield, producer gas composition, heating value of producer gas and cold gas efficiency. Additionally, the temperature profile along the height of gasifier was also investigated.

3.1 Feedstock and feedstock preparation

Feedstock used in this study is PEFB collected from the palm oil mill in Chonburi province which is located in Eastern part of Thailand. After collecting process, PEFB is prepared before being used as feedstock. The important properties of solid fuel for using as feedstock in gasification process are proximate analysis, ultimate analysis, bulk density and heating value as shown in Table 7. The proximate analysis was determined according to ASTM D 5142, the ultimate analysis, e.g. carbon, hydrogen and nitrogen was investigated using ASTM D 5373 and sulfur containing in PEFB was analyzed by ASTM D 4239 (Miller & Tillman, 2008). Because of the complexity in determining oxygen directly, it was determined by difference, i.e. subtracting the total percentage of carbon, hydrogen, nitrogen and sulfur. The heating value was determined by an adiabatic bomb calorimeter as described in ASTM D 5865 (Miller & Tillman, 2008).

	Unit	Value
Proximate analysis		
Moisture (as received after solar drying)	% wt.	8.34
Volatile matter	% wt.	79.82
Fix carbon	% wt.	13.31
Ash	% wt.	6.87
Ultimate analysis		
C	% wt.	43.80
H	% wt.	6.20
O	% wt.	42.65
N	% wt.	0.44
S	% wt.	0.09
Other properties		
Bulk density	kg/m^3	112.04
Lower Heating Value	MJ/kg	19.25

Table 7. Proximate analysis, ultimate analysis and other properties of PEFB (dry basis)

Regarding to the proximate analysis, the volatile matter of PEFB is rather high with the value of 79.82 %. Volatile matter indicates the portion driven off in gas or vapor form which comprises mainly hydrogen, oxygen, carbon monoxide, methane and other hydrocarbons (Miller & Tillman, 2008). The use of fuel with high volatile matter results in the low combustion temperature because some parts of heat is used to vaporize volatile matters in fuel. PEFB contains 8.34 % wt. moisture, 13.31 % wt. fixed carbon and 6.87 % wt. ash. The amount of fixed carbon represents the combustible residue after driving off the volatile matter (Miller & Tillman, 2008) and plays an important role on the amount of CO produced in the reduction zone, which is the main composition of producer gas. From the ultimate analysis, PEFB contains 43.80 % wt. carbon and 6.20 % wt. hydrogen. Both carbon and hydrogen effect the thermo-chemical conversion and then the producer gas composition. PEFB has low sulfur content of about 0.09 % wt. which indicates the tendency of SO_2 and H_2S formation. Other important parameters are bulk density and heating value which affect the gasification behavior and also the quality of producer gas.

In this study both *as received PEFB* and *pelletized PEFB* were used as feedstock. In order to prepare feedstock, as received PEFB was solar dried and cut into small size of 2 cm x 5 cm x 5 cm by cutting machine. In case of using pelletized PEFB as feedstock, pelletizing machine is used for preparing feedstock and the pelletized PEFB has a final diameter of approximately 5.5 cm and a length of 6 cm, as illustrated in Figure 9. The physical properties of pelletized PEFB compared to as received PEFB is listed in Table 8.

(a)   (b)

Fig. 9. PEFB before (a) and after (b) pelletizing

Physical properties	Before pelletizing	After pelletizing
Dimension (cm)	Width 2 cm Length 5 cm Height 5 cm	Diameter 5.5 cm Length 6 cm
Density (kg/m³)	112.04	293.65
Moisture content (%)	8.34	4.9

Table 8. Physical properties of pelletizing PEFB

From Table 8, it can be obviously observed that after pelletizing PEFB is denser and has higher density than as received PEFB because bulk PEFB is pressed and its structural bound attached to each other. This is the advantages of pelletized PEFB because it is cheap to handle, transport and store. Apart from handling and storing behavior, the bulk density is important for the performance of biomass gasification in fixed bed reactor (Knoef, 2005). During the pelletizing process some water containing in PEFB is driven out. These can lead to the decrease in moisture content of PEFB.

3.2 Experimental setup

A 10 kg/hr lab-scale downdraft gasifier which is belonging to the Waste Incineration Research Center (WIRC) and located at Department of Mechanical and Aerospace Engineering, King Mongkut's University of Technology North Bangkok, Thailand is a vertical reactor with fuel feeding system. The reactor is 2,000 mm height and has a diameter of 600 mm. It can be separated into 4 parts as followed: fuel hopper, pyrolysis chamber, reaction chamber and ash chamber, as shown in Figure 10. The temperature in pyrolysis chamber ranges between 200-500 °C, whereas the reaction chamber has the temperature of 500-1200 °C. The temperature in ash chamber is about 300-1000 °C. There are totally 11 type-K thermocouples installed over the height of gasifier: 5 thermocouples in pyrolysis chamber, 4 thermocouples in reaction chamber and 2 thermocouples in ash chamber.

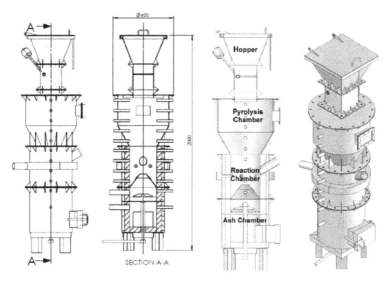

Fig. 10. A 10 kg/hr lab-scale downdraft gasifier

In additional to a downdraft gasifier which is the core component, a lab-scale gasifier system consists of air blower, air pre-heater, gas cleaning unit, weighing apparatus and data logger. Figure 11 shows the process diagram of a lab-scale gasifier system.

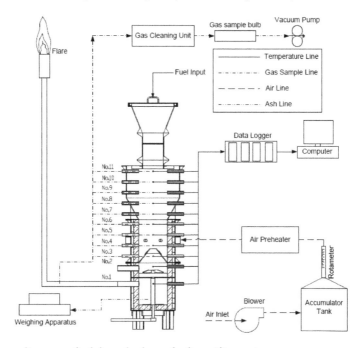

Fig. 11. Process diagram of a lab-scale downdraft gasifier system

3.3 Experiment procedure

In case of as-received PEFB, 2 to 3 kg of PEFB was fed into a 10 kg/hr downdraft gasifier per batch. After feedstock feeding, the lid at the top of gasifier was closed and the feedstock inside the gasifier was ignited by a burner. When feedstock started to be ignited, the gasification air was introduced into gasifier and the gasification process was taken place. When the first batch of feedstock was almost completely gasified, 2-3 kg of feedstock was then introduced again into the gasifier. This step was repeated until the total amount of approximately 25 kg PEFB was fed into gasifier in the whole period of time. In case of pelletized PEFB, 25 kg of it was fed into hopper in one time. During the gasification process, the temperatures inside the gasifier at each position were continuously recorded. If the gasification temperature reached the constant value, the volume flow rate of producer gas were measured. A little amount of producer gas was taken out as gas sample in order to further be investigated, while the remaining producer gas was flared at the stack outlet. The gasification process terminated, when the fuel was completely burnt and the reactor was naturally cooled down. The ash remaining in the reactor was then taken from the reactor and measured the weight in order to determine the percentage of ash production. The air flow rates of 6, 9, 12, 15 and 18 Nm³/hr for as-received PEFB gasification and the air flow rate of 15, 18, 21, 27 and 33 Nm³/hr for pelletized PEFB gasification were varied for each experiment.

After each experiment, producer gas composition, in which only H_2, CO, CO_2, CH_4, N_2 and O_2 were taken into account, was investigated by gas chromatography according to ASTM. Lower heating value of producer gas was also calculated according to Equation 1 and cold gas efficiency was determined by Equation 2.

$$LHV_G = \Sigma v_i \cdot LHV_i \tag{1}$$

$$\eta = \frac{LHV_G \times \dot{V}_G}{LHV_F \times \dot{m}_F} \tag{2}$$

LHV is the lower heating value. The subscript G, i and F refers to the producer gas, each combustible gas component and PEFB, respectively. V_i is the fraction of each combustible gas component in producer gas by volume. $\dot{V}_G$ and $\dot{m}_F$ are producer gas yield by volume and PEFB consumption rate by mass, respectively.

3.4 Results and discussions
3.4.1 Producer gas composition and its lower heating value

The composition of producer gas obtained from air gasification of both as received PEFB and pelletized PEFB is shown in Figure 12 and Figure 13, respectively.

From Figure 12, it can be seen that the concentration of CO increases with increasing air flow rate and its increasing rate is slow down at the higher air flow rate. The concentration of CO_2 decreases until the air flow rate of 9 Nm³/hr and further increases with the air flow rate. H_2-concentration is very fluctuated and cannot predict its tendency from Figure 12, whereas there is no significant change in the concentration of CH_4 for all air flow rates.

Compared Figure 13 to Figure 12, it can be noticed that the concentration of each gas composition is not fluctuated. The tendency of each gas can be predicted from Figure 13. Due to the high density of pelletized PEFB, the fuel is more homogenous and the fuel flow is more stable. Consequently, the reactions between air and fuel during gasification process are more stable and can reach their equilibriums. In case of pelletized PEFB (Figure 13), the concentration of CO and H_2 increases with increasing air flow rate. The increasing rate is more rapid at the air flow rate until 21 Nm^3/hr and for further increase in air flow rate from 21 Nm^3/hr, the concentration of both gases increases slowly or almost remains constant. In contrast to H_2 and CO, the concentration of CO_2 decreases with the air flow rate until its minimum point at the air flow rate of 21 Nm^3/hr. With further increase in air flow rate, CO_2-concentration increases. The concentration of CH_4 in case of pelletized PEFB is almost constant.

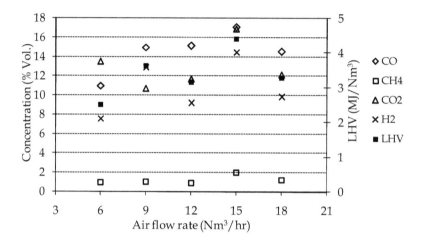

Fig. 12. Producer gas composition with different air flow rates for as-received PEFB

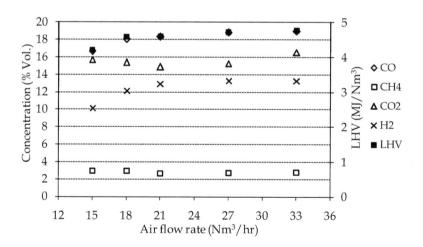

Fig. 13. Producer gas composition with different air flow rates for pelletized PEFB

Considered the reactions occurred in a downdraft gasifier, PEFB is firstly dried and the moisture containing in PEFB is driven off as steam. During pyrolysis process, PEFB is thermally decomposed into gaseous products, tars and chars, as written in Equation 3. Tar which is heavy hydrocarbon compound is also thermally cracked into light hydrocarbons and other gases, as written in Equation 4 (Rui et al., 2007).

$$PEFB \rightarrow gas + tar + char \tag{3}$$

$$Tars \rightarrow light\ and\ heavy\ hydrocarbons + CO + CO_2 + H_2 \tag{4}$$

Gases and the remaining PEFB pass through the oxidation zone where oxidation process occurs. In this zone, combustible gas and combustible material are oxidized to be steam and CO_2 by oxygen containing in gasification air. Equation 5 to Equation 7 shows the examples of oxidation process (Kaltschmitt & Hartmann, 2001; Schmitz, 2001). As the air flow rate increases, the oxidation process is accelerated by increasing amount of O_2 in gasification air and results in the higher reaction temperature (exothermic reactions).

$$H_2 + \frac{1}{2}O_2 \leftrightarrow H_2O \quad \Delta H = -241.8\ kJ/mol \tag{5}$$

$$CO + \frac{1}{2}O_2 \leftrightarrow CO_2 \quad \Delta H = -283.0\ kJ/mol \tag{6}$$

$$C_mH_n + \left(m + \frac{n}{2}\right)O_2 \leftrightarrow mCO_2 + \frac{n}{2}H_2O \tag{7}$$

With further increase in air flow rate, the reactions almost approach their equilibriums. therefore the concentration of each gas composition remains constant. The products of oxidation process react further with other gases and un-reacted fuel in reduction zone. The increase or decrease in composition of producer gas is resulted from reactions in this zone. The increase in CO and H_2 from the experiments is resulted from endothermal Boudouard reaction (Equation 8) and endothermal heterogeneous water gas shift reaction (Equation 9) (Kaltschmitt & Hartmann, 2001; Laohalidanond, 2008; Schmitz, 2001).

$$C + CO_2 \leftrightarrow 2CO \quad \Delta H = 159.9\ kJ/mol \tag{8}$$

$$C + H_2O \leftrightarrow CO + H_2 \quad \Delta H = 118.5\ kJ/mol \tag{9}$$

With increasing air flow rate, the gasification temperature raises as a result of exothermal oxidation. The endothermal Boudouard reaction and endothermal heterogeneous water gas shift reaction are then shifted to the right hand side, consequently, CO and H_2 in producer gas increase. The above mentioned reactions take also the responsibility for the decrease in CO_2 concentration in producer gas.

With respect to the heating value of producer gas, the lower heating value of producer gas yields from as received PEFB is fluctuated and the tendency cannot be predicted because of the non-equilibrium reactions. Taken the results from gasification process of pelletized PEFB into account, it can be remarkably seen that the heating value of producer gas varies with the air flow rate. At the air flow rate of 15 Nm³/hr, the producer gas has the lower heating value of 4.20±0.31 MJ/Nm³ and the lower heating value increases to 4.77±0.29 MJ/Nm³ at the air flow rate of 33 Nm³/hr. The increase in the lower heating value is resulted from the increase in combustible gases, e.g. H_2 and CO with increasing air flow rate.

From the experiments with both as received PEFB and pelletized PEFB, it can be concluded that using pelletized PEFB can provide more stable gasification process than using as-received PEFB and the relevant reactions can approach their equilibriums; hence, pelletized PEFB is more proper to be used as fuel in gasification process than as-received PEFB. Since the producer gas will further be used as fuel in a combustion engine generator for electricity production, the heating value of producer is the major parameter to be concerned. The maximum heating value of 4.77±0.29 MJ/Nm³ is achieved from gasification of pelletized PEFB at the air flow rate of 33 Nm³/hr.

3.4.2 Overall results

This section shows the overall results of experiments with pelletized PEFB at different air flow rates in term of producer gas yield, feedstock consumption rate and cold gas efficiency. Table 9 presents the overall results for pelletized PEFB at different air flow rates.

Air flow rate (Nm³/hr)	Producer gas yield (Nm³/hr)	Fuel consumption rate (kg/hr)	Lower heating value (MJ/Nm³)	Cold gas efficiency (%)
15	24.50	10.74	4.20±0.31	49.71
18	28.70	12.24	4.58±0.29	55.76
21	34.80	15.65	4.60±0.26	53.15
27	41.24	21.64	4.73±0.55	46.80
33	48.89	26.66	4.77±0.29	45.42

Table 9. Overall results for pelletized PEFB at different air flow rates

From Table 9, although the maximum heating value of 4.77±0.29 MJ/Nm³ is taken place at the air flow rate of 33 Nm³/hr, the maximum cold gas efficiency of 55.76 % occurs at the air flow rate of 18 Nm³/hr. In additional to heating value, the ratio of producer gas yield to fuel consumption plays an important role on cold gas efficiency, as clearly seen from Equation 2. At the air flow rate of 33 Nm³/hr, 1 kg of pelletized PEFB can produce 1.83 Nm³ of producer gas, whereas 1 kg of pelletized PEFB can produce 2.34 Nm³/hr of producer gas at the air flow rate of 18 Nm³/hr which is the condition that the maximum producer gas can be yielded from 1 kg of feedstock.

3.4.3 Temperature distribution

To identify the temperature distribution in each reaction zone, the gasification process of pelletized PEFB is only investigated and the result is shown in Figure 14.

The reaction zone for the experiments with different air flow rates is almost identical. Drying zone for moisture removal taking place at the top of gasifier (a height of 70-80 cm) has the temperature of less than 200 °C for all air flow rates. Next reaction zone is pyrolysis zone, 50-70 cm high, which has the pyrolysis temperature of 200-600 °C for all air flow rates. At the height of 30-50 cm, where air is introduced into gasifier, the combustion process occurs and the combustion temperature is 600-1000 °C. At the bottom of a downdraft gasifier (10-30 cm), where the reduction process is taken place, the temperature in the reduction zone is considerable reduced to 400-800 °C.

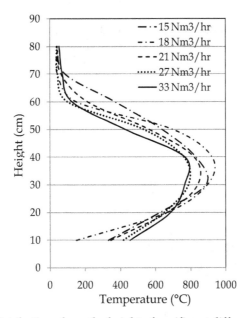

Fig. 14. Temperature distribution along the height of gasifier at different air flow rates

3.5 Conclusion
From the experiments in a laboratory scale downdraft gasifier, it can be implied that both as received and pelletized PEFB has a potential to be used as fuel for producer gas production. However, pelletized PEFB is more suitable than as received PEFB because their reactions in gasification process are more stable and can approach equilibrium. The producer gas obtained from gasification of pelletized PEFB at the air flow rate of 33 Nm³/hr which is the most suitable operating condition consists of 19.02 % wt. CO, 13.32 % wt. H_2, 2.78 % wt. CH_4 and 16.58 % wt. CO_2. It heating value of 4.77 MJ/Nm³ can be achieved with the cold gas efficiency of 45.42 %. The reaction temperature has been classified on 4 different zones; less than 200 °C for drying zone, 200-600 °C for pyrolysis zone, 600-1000 °C for oxidation zone and 400-800 °C for reduction zone.

4. Experimental study of PEFB gasification in a protoype-scale downdraft gasifier

After it is proven that pelleitzed PEFB has a high potential to be used as fuel for producer gas production via gasification process in the previous section, this section aims to conduct the feasibility of using pelletized PEFB as fuel for power generation via gasification process.

4.1 Feedstock and feedstock preparation
Feedstock used for the experiments in a prototype scale gasifier is pelletized PEFB which was prepared by the same method as described in section 3.1. The proximate and ultimate analyses of feedstock as well as other physical properties have already been shown in Table 7 and Table 8, respectively.

4.2 Experimental Setup

A 50 kg/hr prototype downdraft gasification plant is also belonging to WIRC and located in Saha Pathana Industrial Park in Kabin Buri, Prachinburi province, Thailand. This plant consists mainly of 5 parts, as in the following: fuel preparation system, downdraft gasifier, heat exchanger, gas cleaning unit and internal combustion engine-generator, as shown in Figure 15.

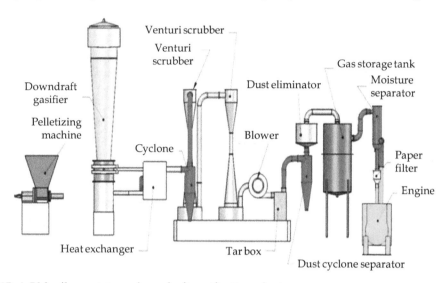

Fig. 15. A 50 kg/hr prototype downdraft gasification plant

The downdraft gasifier can be divided into fuel hopper, reaction zone and ash discharging zone and has a capacity to load fuel about 50 kg/hr for 7 hours operation continuously. Shell and tube heat exchanger is applied in order to preheat the air which is served as gasification agent. Gas cleaning system consists of cyclone, venturi scrubber, and dust removal unit. The internal combustion engine-generator for power generation is a 4-strokes diesel engine with 4- cylinders and can produce 50 kW electricity, 380/400 V and 50/60 Hz, as illustrated in Figure 16.

Fig. 16. Internal combustion engine-generator for power generation

4.3 Experiment procedure

The experiment prodecure is as same as described in Section 3.3 of this chapter but only pelletized PEFB was used as feedstock. The air flow rate for the experiments in a prototype downdraft gasifier was varied from 90 to 120 Nm³/hr with an interval of 10 Nm³/hr. Distinguish from Section 3.3 was that the producer gas obtained from a prototype downdraft gasifier at the most suitable condition was further used as fuel in the internal combustion engine-generator. The electrical load, in this case: electrical heater, varied from 18 to 36 kW with the step of 6 kW. The consumption of both diesel and producer gas was recorded and the rate of diesel replaced by producer gas in percent can be calculated by Equation 10. Finally, the overall efficiency for power production from pelletizing PEFB is calculated by Equation 11.

$$R = \frac{\left(\dot{m}_{d,o} - \dot{m}_{d,d}\right)}{\dot{m}_{d,o}} \tag{10}$$

$$\eta_T = \eta_E \cdot \eta_G \tag{11}$$

Where R is the rate of diesel replaced by producer gas, $\dot{m}_{d,o}$ is the mass flow rate of diesel consumption in case of using diesel as single fuel and $\dot{m}_{d,d}$ is the mass flow rate of diesel consumption in case of using dual fuel. η_T, η_E and η_G represent the overall efficiency, engine efficiency and cold gas efficiency, respectively.

4.4 Results and discussions
4.4.1 Producer gas composition and its lower heating value

The composition and heating value of producer gas obtained from air gasification of pelletized PEFB in a protoype downdraft gasifier are shown in Figure 17.
From Figure 17, the concentration of H_2, CO and CH_4 increases with increasing air flow rate until the air flow rate of approximately 100-105 Nm³/hr and decreases with higher air flow rate. At the air flow rate of 100 Nm³/hr, the highest concentration of CO and CH_4 can be obtained, consequently, the maximum heating value of 6.99 MJ/Nm³ also occurs.

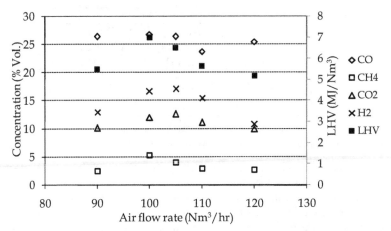

Fig. 17. Producer gas composition with different air flow rates for pelletized PEFB in prototype gasifier

4.4.2 Overall results

Table 10 shows the overall results for pelletized PEFB at different air flow rates, in term of producer gas yield, fuel consumption rate and cold gas efficiency.

Air flow rate (Nm³/hr)	Producer gas yield (Nm³/hr)	Fuel consumption rate (kg/hr)	Lower heating value (MJ/Nm³)	Cold gas efficiency (%)
90	111.40	45.08	5.49±0.10	70.47
100	129.44	61.57	6.99±0.29	76.34
105	176.73	74.82	6.50±0.15	79.73
110	182.19	80.58	5.64±0.23	66.21
120	191.76	88.09	5.18±0.40	58.55

Table 10. Overall results for pelletized PEFB at different air flow rates in prototype gasifer

From Table 10, the maximum heating value of 6.99±0.29 MJ/Nm³ is observed at the air flow rate of 100 Nm³/hr which correspondances to the cold gas efficiency of 76.34 %. From the observation during the experiments, although at the air flow rate of 100 or 105 Nm³/hr the heating value of producer gas and the cold gas efficiency reach the maximum value, the producer gas was unstable and non-continuously formed. Instead, at the air flow rate of 110 Nm³/hr, the producer gas was continuously and uniformly generated. Therefore, for testing of using producer gas as fuel in the internal combustion engine-generator, the air flow rate of 110 Nm³/hr is selected as optimum operating condition.

4.4.3 Testing of producer gas in an internal combustion engine-generator

Table 11 shows the results of testing of producer gas in the internal combustion engine-generator with the electrical loads of 18, 24, 30 and 36 kW. At the beginning, only diesel fuel was used as fuel in order to examine the diesel consumption rate. Later, the experiments of using dual fuel, in this case: diesel fuel and producer gas, were carried out.

Load (kW)	Diesel fuel	Dual fuel			
	Diesel consumption (kg/hr)	Producer gas consumption (Nm³/hr)	Diesel consumption (kg/hr)	Diesel replacing rate (%-wt.)	Engine efficiency (%)
18	5.41	37.63	2.49	53.97	20.01
24	6.44	53.60	1.05	83.69	24.72
30	7.78	53.60	2.19	71.85	26.96
36	9.12	39.91	4.32	52.63	30.95

Table 11. Results for testing of producer gas in the internal combustion engine-generator

It can be noticed that the producer gas consumption increases with increasing electrical load until the electrical load of 30 kW and for the higher electrical load, the producer gas consumption decreases. It means the producer gas can replace diesel fuel successfully if the electrical load is increased but at the higher electrical load, the producer gas cannot further replace diesel fuel due to the low heating value of producer gas which is not sufficient to sustain the higher load. The maximum diesel replacing rate of 83.69 % is taken place at the electrical load of 24 kW, by which the engine efficiency is accounted to be 24.72 %. At this point, the overall efficiency for power generation from pelletizing PEFB via gasification process is calculated to be 16.36 %.

4.5 Conclusion

From the experiments in a prototype downdraft gasifier, it can be concluded that the producer gas obtained from pelletized PEFB can be used as a substitute fuel to conventional diesel fuel. The optimum air flow rate for gasification process is 110 Nm^3/hr, by which the producer gas was continuously and uniformly generated. The producer gas contains 23.74 % wt. CO, 15.48 % wt. H_2, 2.97 % wt. CH_4 and 10.01 % wt. CO_2. The heating value is 5.64±0.23 MJ/Nm^3 and the cold gas efficiency is 66.21 %. After using this producer gas in an internal combustion engine-generator, it can be found that the diesel fuel consumption can be reduced by more than 80 % at the electrical load of about half-load (24 kW) and the overall efficiency of 16.36 % can be achieved at this load.

5. General conclusion

As Thailand is an agricultural base country, there are a lot of agricultural residues left over. These residues can be used as alternative fuel to replace the conventional fuel which needs to be imported from foreign countries. PEFB is one of the most available agricultural residues generated from palm oil industry. From this study, it can be found that PEFB, especially pelletized PEFB, has a very high potential to be used as alternative fuel for power production via gasification process. The producer gas obtained from a laboratory scale downdraft gasifier at the air flow rate of 33 Nm^3/hr consists of 19.02 % wt. CO, 13.32 % wt. H_2, 2.78 % wt. CH_4 and 16.58 % wt. CO_2. Its heating value of 4.77 MJ/Nm^3 can be achieved with the cold gas efficiency of 45.42 %. The reaction temperature has been classified on 4 different zones; less than 200 °C for drying zone, 200-600 °C for pyrolysis zone, 600-1000 °C for oxidation zone and 400-800 °C for reduction zone. The producer gas obtained from a prototype scale downdraft gasifier posses a very high heating value varied from 5.18-6.99 MJ/Nm^3 depending on the air flow rate. At the optimum air flow rate of 110 Nm^3/hr, the producer gas contains 23.74 % wt. CO, 15.48 % wt. H_2, 2.97 % wt. CH_4 and 10.01 % wt. CO_2. The heating value is 5.64 MJ/Nm^3 which can effectively replace the diesel consumption in the internal combustion engine-generator. The diesel replacement rate of more than 80 % can be obtained at the electrical load of 24 kW and the overall efficiency is 16.36 %. From this study, it can be concluded that PEFB can be used as alternative fuel for heat or electricity production, for eco-friendly and sustainable development in Thailand.

6. Acknowledgements

The authors would like to appreciate Faculty of Engineering and Science Technology Research Institute of King Mongkut's University of Technology for the financial support.

7. References

Abdullah, N.; Bridgwater, A.V. (2006), Pyrolysis Liquid Derived from Oil Palm Empty Fruit Bunches, *Journal of Physical Science*, Vol.17, No.2, pp. 117-129, ISSN 1992-1950

Abdullah, N.; Gerhauser, H.; Bridgwater, A.V. (2007), Bio-oil from Fast Pyrolysis of Oil Palm Empty Fruit Bunches, *Journal of Physical Science*, Vol.18, No.1, pp. 57-74, ISSN 1992-1950

Amin, N.A.S.; Asmadi, M. (2008), Optimization of Empty Palm Fruit Bunch Pyrolysis over HZSM-5 Catalyst for Production of Bio-oil, In: Universiti Tecknologi Malaysia, 5 March 2011, Available from: http://eprints.utm.my/5125

Azizan, M.T.; Yusup, S.; Laziz, F.D.M.; Ahmad, M.M. (2009), Production of Bio-oil from Oil Palm's Empty Fruit Bunch via Pyrolysis, *Proceedings of the 3rd WSEAS International Conference on Renewable Energy Sources*, ISBN 978-960-474-093-2, Spain, July 2009

Belgiorno, V.; De Feo, G.; Della Rocca C.; Napoli, R.M.A. (2003), Energy from Gasification of Solid Wastes, Waste Management, Vol.23, No.1, (January, 2003), pp. 1-15, ISSN 0956-053x

BP (2010), BP Statistical Review of World Energy June 2010, Available from: www.bp.com

Bull S.R. (2001), Renewable Energy Today and Tomorrow, *Proceedings of the IEEE*, Vol.89, No.8, (August 2001), pp. 1216-1226, ISSN 0018-9219

Department of Provincial administration (2010), Population Statistic of Thailand, In: *Department of Provincial administration*, May 2011, Available from: www.dopa.go.th

Energy Policy and Planning Office (2010), Energy Statistic of Thailand 2010, In: Energy Policy and Planning Office, 25 January 2011, Available from: www.eppo.go.th/info/cd-2010/index.html

Hamzah, M.M.B. (2008), The production of ecofiber from palm oil empty fruit bunch, In: *Universiti Malaysia Pahang*, 5 February 2011, Available from: http://umpir.ump.edu.my/521

Heriansyah (n.d.), Optimizing the use of oil palm by-product (EFB) as fertilizer supplement for oil palm, In: *BW Plantation*, 2 September 2010, available from: http://www.bwplantation.com/document/Optimizing%20The%20Use%20of%20Empty%20Fruit%20Bunch%20(EFB).pdf

Hussain, A.; Ani, F.N.; Darus, A.N.; Mokhtar, H.; Azam, S.; Mustafa, A. (2006), Thermo-chemical Behaviour of Empty Fruit Bunches and Oil Palm Shell Waste in a Circulating Fluidized-Bed Combustor (CFBC), *Journal of Oil Palm Research*, Vol.18, No. 1, (June 2006), pp. 210-218, ISSN 1511-2780

Kaltschmitt, M. ; Hartmann, H. (2001), *Energie aus Biomasse – Grunglagen, Techniken und Verfahren*, VDI-Verlag, ISBN 978-354-064-8536, Germany

Katamanee, A. (2006), Appropriate technology evaluation for oil palm product utilization in Krabi province, In: *Mahidol University*, 1 February 2011, Available from: www.li.mahidol.ac.th/thesis/2549/cd388/4637145.pdf

Knoef, H.A.M. (2005), Practical aspects of biomass gasification, In: *Handbook of biomass gasification*, BTG biomass technology group BV, pp. , Druckerij Giehoorn ten Brink, ISBN 978-908-100-6811, Natherlands

Laohalidanond, K. (2008), *Theoretische Untersuchungen und thermodynamische Modellierungen der Biomassevergasung und der Fischer-Tropsch-Synthese zur Herstellung von Diesel kohlenwasserstoffen aus thailaendischen Biomassen*, Shaker Verlag, ISBN 978-383-227-8250, Germany

Law, K.N.; Daud, W.R.W.; Ghazali, A. (2007), Morphological and Chemical Nature of Fibre Strands of Oil Palm Empty Fruit Bunch (OPEFB), *Bioresources*, Vol.2, No.3, pp. 351-362, ISSN 1930-2126

Mahlia, T.M.I.; Abdulmuin, M.Z.; Alamsyah, T.M.I.; Mukhlishien, D. (2001), An alternative energy source from palm waste industry for Malaysia and Indonesia, *Energy Conversion and Management*, Vol.42, No.18, (December 2001), pp. 2109-2118, ISSN 0196-8904

Miller, B.G. ; Tillman, D.A. (2008), Coal Characteristics, In: *Combustion engineering issues for solid fuel systems*, Miller, B.G. ; Tillman, D.A., Elsevier, pp. 33-82, ISBN978-012-373-6116, United State of America

Nasrin, A.B.; Ma, A.N.; Choo, Y.M.; Mohamad, S.; Rohaya, M.H.; Azali, A.; Zainal, Z. (2008), Oil Plam Biomass As Potential Substitution Raw Materials For Commercial Biomass Briquettes Production, *American Journal of Applied Science*, Vol.5, No.3, pp. 179-183, ISSN 1554-3641

Obernberger, I.; Thek G. (2008), Combustion and gasification of solid biomass for heat and power production in Europe state-of-the-art and relevant future development, *Proceedings of the 8th European Conference on Industrial Furnaces and Boilers*, ISBN 978-972-993-0935, Portugal, April, 2008

Office of Agricultural Economic (2010), In: *Agricultural Statistical Data*, 20 December 2010, Available from: www.oae.go.th

Office of the National Economic and Social Development Board (2009), National income of Thailand 2009 edition, In: *Office of the National Economic and Social Development Board*, 29 June 2011, Available from: www.nesdb.go.th

Owen A.D. (2006), Renewable energy: Externality costs as market barriers, *Energy Policy*, Vol.34, No., (), pp. 632–642, ISSN 0301-4215

Papong, S.; Yuvaniyama, C.; Lohsomboon, P.; Malakul, P. (2004), Overview of Biomass Utilization in Thailand, *ASEAN Biomass Meeting*, Bangkok, 2004

Perez, R. (1997), Feeding pigs in the tropics, In: *Food and Agriculture Organization of the United Nations Rome*, 1 February 2011, Available from: www.fao.org/docrep/003/w3647e/W3647E00.htm

Prasad, B.V.R.K.; Kuester, J.L. (1998), Process analysis of a dual fluidized bed biomass gasification system. *Journal of Industrail & Engineering Chemistry Research*, Vol.27, No.2, (February 1988), pp. 304-310, ISSN 0888-5885

Prasertsan, S.; Sajjakulnukit, B. (2006), Biomass and biogas energy in Thailand: Potential, opportunity and barriers, *Renewable Energy*, Vol.31, No.5, (April 2006), pp. 599-610, ISSN 0960-1481

Ramli, R.; Shaler, S.; Jamaludin, M.A. (2002), Properties of medium density fibreboard from oil palm empty fruit bunch fibre, *Journal of Oil Palm Research*, Vol.14, No. 2, (December 2002), pp. 34-40, ISSN 1511-2780

Renewable Energy Policy Network for the 21st Century (2009), Renewables Global Status Report 2009 update, In: *Renewable Energy Policy Network for the 21st Century*, 13 December 2010, Available from: http://www.ren21.net/pdf/RE_GSR_2009_update.pdf

Rui, X.; Baosheng, J.; Hongcang, Z.; Zhaoping, Z.; Mingyao, Z. (2007), Air gasification of polypropylene plastic waste in fluidized bed gasifier, *Energy Conversion and Management*, Vol.48, No.3, (March 2007), pp. 778-786, ISSN 0196- 8904

Sawangphol, N.; Pharino, C. (2011). Status and outlook for Thailand's low carbon electricity development, *Renewable and Sustainable Energy Reviews*, Vol.15, No.1, (January 2011), pp. 564-573, ISSN 1364-0321

Schmitz, W. (2001), *Konversion biogener Brennstoffe fuer die Nutzung in Gastubinen*, VDI-Verlag, ISBN 978-318-345-9063, Germany

Shuit, S.H.; Tan, K.T.; Lee, K.T.; Kamaruddin, A.H. (2009), Oil Palm Biomass as a sustainable energy source: A Malaysian Case Study, *Energy*, Vol.34, No.9, (September 2009), pp. 1225-1235, ISSN 0360-5442

Sukiran, M.A.; Chin, C.M.; Baker, N.K.A. (2009), Bio-oil from Pyrolysis of Oil Palm Empty Fruit Bunches, *American Journal of Applied Science*, Vol.6, No.5, pp. 869-875, ISSN 1546-9239

U.S. Energy Information Administration, 2006. World Carbon Dioxide Emissions from the Consumption and Flaring of Fossil Fuels 1980-2006, In: *International Energy Annual 2006*, 30 April 2011, Available from: www.eia.gov/iea/carbon.html

U.S. Energy Information Administration, 2010. Energy-related carbon dioxide emission, In: *International Energy Outlook 2009*, 30 April 2011, Available from : www.eia.doe.gov/oiaf/ieo/emissions.html

Wainwiwat, S.; Asafu-Adjaye, J. (2011). Modelling the promotion of biomass use: A case study of Thailand, *Energy*, Vol.36, No.3, (March 2011), pp. 1735-1748, ISSN 0360-5442

Yang, H.; Yan, R.; Chen, H.; Lee, D.H.; Liang, D.T.; Zheng, C. (2006), Mechanism of Palm Oil Waste Pyrolysis in a Packed Bed, *Energy & Fuels*, Vol.20, No.3, (April 2006), pp. 1321-1328, ISSN 0887-0624

Yangdee, B. (n.d.), Ten million rai of oil palm plantation: A catastrophe for the Thai people, In: *World rainforest movement*, 1 February 2011, Available from: www.wrm.org.uy/countries/Thailand/Catastrophe.pdf

Yusoff, S. (2006), Renewable energy from palm oil innovation on effective utilization of waste, *Journal of Cleaner Production*, Vol. 14, No.1, pp. 87-93, ISSN 0959-6526

6

Renewable Energy Sources vs Control of Slovak Electric Power System

Juraj Altus, Michal Pokorný and Peter Braciník
University of Žilina,
Slovak Republic

1. Introduction

The chapter presents issues of renewable energy sources connection into Slovak power system. It describes calculation techniques for estimating acceptable RES capacities, which can be connected into the system, without jeopardizing security and quality of supplied electricity. Calculations of additional expenses for regulation power purchase for various values of RES capacities connected are presented in the second part of the chapter.

The interest in construction of photovoltaic power plants in Slovak Republic soared in 2008. A new act concerning renewable energy sources and remarkably generous prices paid for electricity from these sources (associated with the act) stimulated investors' interest in building new plants. Based upon experience from other countries and following analysis of possibilities for procuring own sources of balancing electricity for auxiliary services, Slovak Electricity Transmission System, Plc. (SEPS) placed an order to Department of Power Electrical Systems to elaborate a study in which a maximum capacity of PV plants would be determined, which can be connected into the power system, taking into account availability of balancing power for secondary and tertiary regulation. Developed study determined 120 MWp as maximum power of PV plants, which can be connected in 2010.

Influence from EU side and from investors and construction of new thermal and nuclear power plants in Slovakia necessitated an elaboration of a new study, which would analyse the situation in transmission system with respect of balancing electricity for the case, when new PV plants with capacity up to 1 200 MWp are connected. This chapter presents solution of this matter and some results.

Determination of connectable capacity of PV plants into power system of Slovakia can be accomplished from different points of view. One of them can be based on the basic condition, which is used in all studies concerned PV plants connection into distribution network. This condition states, that voltage change in point of common coupling should be less than 2 %, compared with the situation before plant connection. If this condition is applied, resultant connectable capacity should be relatively high, as 22 kV lines can easily withstand considerably higher loads than those of current ones. Probably 80 to 90 % of demanded capacity could be accepted in this case. This consideration can be hardly accepted due to sources features, which show a vast variability of power, practically in whole power range. In addition, capacities of PV plants are usually lower than 4 MW, thus the responsibility for balance is brought to the distribution company. But the power balance can only be solved on the Slovak transmission system level.

The other possible approach is according to the act [1], where in § 1 sect. 2 the following is stated: „Operator of distribution system is obliged, after the price for connection to distribution system is reimbursed, to connect installation of electricity producer into distribution system with priority, if the installation fulfils technological conditions and sales conditions of connection into the system so that security, reliability and stability of system operation is unchanged". Operator of regional distribution system uses electricity obtained according to sect. 6 for losses redemption. In case, when instantaneous power of obtained electricity exceeds the value necessary for losses redemption, operator of regional distribution system has the right to sell this electricity for the market price. This kind of electricity selling is not considered a business activity in energy sector and does not require a license for electricity supply.

The third possible approach for connectable capacity into distribution system determination is based on the assumption of PV sources variability and the responsibility for balance being bear by a distribution company. Purchase of auxiliary services is needed for balance compensation and this purchase can only be accomplished by Slovak transmission system operator. So the approach to PV plants connectivity with regard to the balance responsibility has to be solved on the Slovak transmission system level, cumulative for all PV plants connected into distribution systems in Slovakia. This approach is described in the following sections of the chapter. In this analysis wind power plants are also taken in consideration together with PV plants. Their connectable capacity was determined to maximum 200 MW. Generally, the term „renewable energy sources (RES)" is used in the chapter.

2. Input data for PV plants influence analysis

Slovak Electricity Transmission System, Plc. was established on 1 April 1994. That day a former National Power System Dispatch Centre in Prague finished its activities and Slovak Power System Dispatch Centre in Zilina took over. Slovak power system was step by step transformed into several economically independent units.

Electricity production is concentrated in Slovenske Elektrarne, Plc, a part of ENEL Group (the company is partly owned by Slovak government).

Production sources are in the following structure:

nuclear sources	1820 MW
thermal sources	2584 MW
hydro sources	2478 MW
others	898 MW

Slovak Electricity Transmission System, Plc. with its National Power System Dispatch Centre in Zilina are performing transmission and controlling activities on transmission system.

This company operates:

- 1776 km of 400 kV lines
- 902 km of 220 kV lines
- 17 substations of 400 kV
- 8 substations of 220 kV

Total transformation capacity is 10 010 MVA.

Three distribution companies provide distribution services for end-consumers. These are:

- Zapadoslovenska energetika, Plc. (in Western Slovakia)
- Stredoslovenska energetika, Plc. (in Central Slovakia)
- Vychodoslovenska energetika, Plc. (in Eastern Slovakia)

owned by foreign companies.

The maximum load of the power system in 2010 measured on 17 December 2010, 17:00 h was 4 342 MW.

Minimum load in the same year measured on 8 August 2010, 5:00 h was 2 190 MW.

Today, there is about 470 MW of installed capacity in photovoltaic power plants in Slovak power system. Wind sources capacity is practically negligible. Hydro power capacity was already mentioned. Photovoltaic sources are characterized by rapid changes of powers, depending on weather conditions. Despite the fact that prediction of performance of RES is in progress these sources are causing unbalance between electricity production and consumption – mainly during periods of low loads in system. The main problem is the possibility to build PV plants with the capacity up to 1 MW. These plants (after connection into system) are not obliged to control deviation between agreed and actual electricity supply into network. These plants are connected to distribution 22 kV system (in 99 % of cases) and are not obliged to measure parameters and transfer data to dispatching centre. Mentioned deviations in electricity supply have to be handled by dispatching centre through purchasing auxiliary services. These additional purchases increase electricity price for customers. Also quality of supplied electricity can be affected. Different types of auxiliary services are described in the following sections together with analysis of their impact on the Slovak power system operation.

A basic property of power system operation is that equilibrium between production and consumption of electricity has to be maintained in every single moment. The consumption of electricity is given by consumers themselves by switching on and off a large number of different appliances and that is why can hardly be affected. Thus equilibrium has to be maintained on the electricity production side and the sources' power must be adapted to instantaneous consumption.

Power system load planning is based on, considering permanent time changes of electricity consumption, the behaviour of system load during 24 hours depicted in daily load diagram (DLD). Expected load during a year is determined from typical DLDs for power system of two days with the highest and lowest loads. DLD for Slovak power system is depicted in Fig. 1 for a summer day.

Calculations for auxiliary services were done only for summer season in the time of minimal value of weekly maximum - L_{MAX} for the values of 2261, 2394, 2660, 2926, 3192, 3458 MW. Evaluated scenarios of system electricity sources employment participating in coverage of DLD in regulation area of Slovakia during summer, in time of absolute minimum were considered according to available data and records from SEPS. In the calculation of the Slovak power system (SPS) operation substantial changes in electricity production installations can are considered. These are listed in Table 1 for individual years. For the purpose of simplified depiction of sources employment, the scenarios were marked A, B, C.

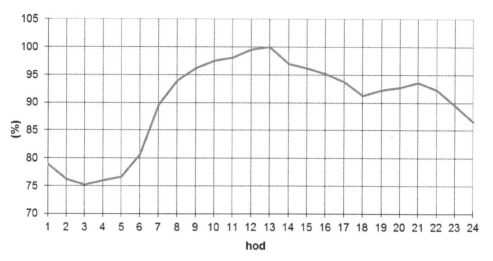

Fig. 1. DLD of summer day

Sources employment scenario		A	B	C
Source/Year		2010	2011	2013
Nuclear power plant	EMO 3,4			1000 MW
Combined cycle plant	Malženice		430 MW	430 MW
	New			850 MW
RES	wind+PV	variable	variable	variable

Table 1. Expected changes of installed capacity for years 2010 – 2013

For each of the sources employment scenarios A, B, C the following installed capacities of RES were used: 300, 400, 500, 600, 800, 1000 and 1200 MW. Each capacity of RES was then employed in the values of 10, 20, 40, 60, 70, 80 and 90 % of installed capacity.

Based on actual operational states of existing RES in SPS the installed capacity utilisation of RES in summer is 0 – 75 % P_{inst} . The variability depends mainly on actual time changes of global solar radiation and used photovoltaic panels.

The coverage of DLD depends on possibilities of each power plant, their failure rate and planned repairs cycles. Employment of individual sources for DLD coverage was performed according to the standards used in preparation of transmission system operation. Pumped storage plants were used to cover peaks of the DLD. Operation of industrial power plants in different regions was considered according to previous years' information. Cut-offs of production facilities and used electricity production technology have substantial impact on source employment in summer season. In case of cut-offs of production facilities, data from SEPS for the years 2010 and 2011 were used.

3. Methodology and calculations of auxiliary services necessary for RES regulation

The purpose of auxiliary services is provision of steady power balance. On one side there is electricity production starting from traditional sources to RES, on the other side are customers, i.e. final consumers. Production and consumption within the scope of interconnected power system must be in equilibrium at every moment.

A new approach to RES support, embodied in act No. 309/2009 about the support of RES and highly effective combined production [4], brings a quasi new group of producers to the electricity market. Those producers produce electricity not according to market demand but practically any time when climatic (wind, solar flux) conditions allow. Responsibility for sales of produced electricity and for the potential balance from the planned values are transferred primarily to operators of distribution systems who are obliged to buy electricity from RES, but eventually power balanced must be maintained by TSO. Supportive mechanisms for RES are currently only starting and massive capacities of these sources are not installed in SPS yet. The progress in installed capacity of RES in distribution system in 2010 is depicted in Fig. 2. Development trend of installed capacity is soaring confirming relevance of topic of availability and sufficiency of regulative capabilities and possibilities of SPS.

3.1 Methodology of setting necessary auxiliary services

The setting of necessary range of auxiliary services for securing reliable operation is closely linked to the degree of reliability of the system. The higher the rate of reliability is required, the higher the range of auxiliary services is needed, having a substantial influence on the final electricity price.

When setting the necessary range of auxiliary services important source information include not only expected loads in regulation area but also load diagram for considered interval of time, value of installed capacity of RES and other statistical data associated with system operation. Amount of auxiliary services was calculated according to the methodology published in [1].

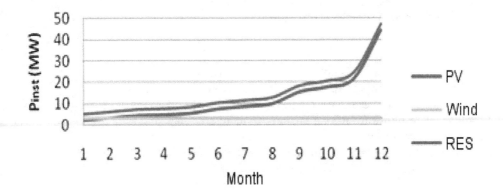

Fig. 2. Progress in installed capacity of RES in 2010

3.2 Setting necessary auxiliary services range for RES regulation

When setting the necessary amount of auxiliary services for SPS needs long term statistics of load and system balance are utilised. Because there is no centralised measurement of power on existing RES, it is not easy to define influence of RES to SPS especially to volumes of auxiliary services. SEPS can only expect that operator of distribution systems will be willing to provide high quality prediction of production from RES and placement of the electricity in the market thus minimising requirements for regulation reserves.

3.3 Impact of powers of RES fluctuations on primary frequency regulation

The role of primary frequency regulation (PRV) is to avoid the occurrence of impermissible deviation in interconnected power system during few seconds. Resulting from the nature of primary frequency regulation the considerable deviations of balance between production and consumption caused by outages of large electricity sources are compensated within seconds. Overall power reserve of 3 000 MW is necessary to secure functionality of the primary regulation in the interconnected international system RG CE [8].
Primary regulation is of proportional nature and maintains equilibrium of production and consumption in synchronous interconnected area based upon frequency deviation. Aliquot part of the primary regulation reserve of Slovakia for 2010 is $PRV = \pm 30$ MW. Value of active power is symmetrical, which means ± 30 MW.

3.4 Impact of powers of RES fluctuations on secondary power regulation

Secondary power regulation (SRV) maintains equilibrium of production and consumption as well as system frequency in each regulation area taking into account regulation programme without violation of primary regulation, that works concurrently in synchronous interconnected area.
Secondary regulation uses centralised automatic production regulation which maintains setting of active power of production units during seconds up to 15 minutes (typically) after the event. Secondary regulation is based on the secondary regulation reserves controlled automatically. Adequate secondary regulation depends on production sources offered by production companies for disposal for auxiliary services.
Minimal recommended value for secondary regulation reserve within interconnected system RG CE is derived from the expected value of maximum system load in give time period according to the empirical formula [8]:

$$SRV_{RGCE} = \pm\sqrt{a \cdot L_{max} + b^2} - b \, , \tag{1}$$

where $a=10$ (empirical constant), $b=150$ (empirical constant),
L_{max} - expected maximum load, SRV - secondary power regulation.
The other part of the secondary regulation reserve within the regulation area Slovakia is a component resulting from load changes dynamics of regulation area ($SRV_{DYN,L}$). The value of this component can be derived from the statistics monitoring system load during a longer period of time and is within the range of 20 to 40 MW, which means circa 1 % recalculated to average yearly load of SPS.

$$SRV_{DYN,L} = \frac{R_\varphi}{2} + \sigma \, , \tag{2}$$

where R_ϕ – arithmetic average of 10 minute differences of maximal and minimal load values for whole hours,

σ – standard deviation.

Resultant value of secondary regulation of regulation area Slovakia then equals to sum of minimal recommended value RG CE and a dynamic component. The value of power is symmetrical.

$$SRV_{FIN} = \pm (SRV_{RGCE} + SRV_{DYN,L}) \tag{3}$$

With the help of secondary regulation the central controller of regulation area maintains compensation of frequency deviations and compensation of active power balance on the planned level. The value of active power is symmetrical. Minimal offered power for SRV is ± 2 MW per unit. The whole regulation range must be realised within 15 minutes from the request and has to be symmetrical according to the basic power. Basic power is unit power for DLD coverage determined by provider within the preparation of operation.

Unit has to allow continuous repeated power changes in any direction within offered regulation range for SRV. Offered regulation power has to be available during whole negotiated time period (hour, day, etc.).

When calculating required range of secondary power regulation it is necessary to consider the influence of RES mainly wind and PV plants. RES are causing additional power fluctuations is regulation area. This undesirable phenomenon has a direct impact on secondary regulation reserve increase. Various foreign system analyses proved the fact that after implementing RES dynamic variations and increase/decrease gradient of non-covered load (difference between overall load and RES production) have risen. In view of the secondary frequency regulation mission (whose role is to maintain dynamic unbalance between planned production and expected load, and thus to keep the balance of regulation area) a new component of secondary regulation reserve $SRV_{DYN,RES}$ has to be introduced.

Additional components of secondary regulation reserve take into account fluctuation of load and production of RES. As there is neither mutual relation nor dependency between mentioned components, these components cannot be directly arithmetically added. If the arithmetical addition is used the value of overall secondary regulation reserve would rise inadequately. One way how to consider both non-correlating components is the use of the function which calculates geometric sum of the values. Resultant value of secondary regulation reserve is then symmetrical and can be calculated according to the following formula:

$$SRV_{VYS} = \pm (SRV_{RGCE} + \sqrt{SRV_{DYN,L}^2 + SRV_{DYN,RES}^2}) . \tag{4}$$

Component $SRV_{DYN,RES}$ constitutes of two partial components, which consider the influence of wind and PV plants and final value can be calculated according to the following formula:

$$SRV_{DYN,RES} = \pm (SRV_{DYN,WIND} + SRV_{DYN,PV}) . \tag{5}$$

Amounts of secondary regulation reserve for wind plants $SRV_{DYN,WIND}$ were specified according to the findings in study [9]. Amounts of secondary regulation reserve for PV plants $SRV_{DYN,PV}$ were specified according to the statistical values from PV plants in operation in Czech Republic and Slovakia [23, 24]. However it has to be remarked that considered were only roof applications with low installed capacity and mentioned statistical data did not include long term information.

P_{inst} (kW)	10	20,1	62,4	68,3	84
$\Delta P_{hour} \left(\% P_{inst} \right)$	2,8	2,4	1,8	1,7	1,4

Table 2. Hourly changes of PV plants power

The basis for the necessary range of the secondary regulation reserve $SRV_{DYN,PV}$ determination is hourly sample of expected time changes of overall PV plants' power taking into account hourly power shown given in Table 2. Additional values for secondary regulation reserve were calculated according to the method of $SRV_{DYN,PV}$ calculation.

Based on statistics, performed calculations and after adaptation of the values for the conditions in Slovakia the expected volumes of secondary regulation reserve for RES $SRV_{DYN,RES}$ were determined and are shown in Table 3.

$P_{inst\ RES}$ (MW)	300	400	500	600	800	1000	1200
$SRV_{DYN,RES}$	43	50	58	65	80	95	109
$SRV_{DYN,RES}$ (% of $P_{inst\ RES}$)	14,29	12,57	11,53	10,84	9,98	9,46	9,11

Table 3. Expected volumes of secondary regulation reserve for RES

3.5 Impact of powers of RES fluctuations on tertiary power regulation

Tertiary regulation (TRV) uses tertiary reserve, which is usually activated manually by TSO in the case of actual or expected secondary regulation activation.

Tertiary regulation is principally used for secondary reserve's release in the balanced state of the system, but it is also activated as a supplement of the secondary reserve after larger outages for system frequency restoration and following release of primary reserve within the whole system. Tertiary regulation is typically performed within the responsibility of TSO.

The nature of tertiary regulation differs from that of secondary regulation. While secondary regulation maintains dynamic unbalance between planned production and expected consumption, tertiary regulation corrects errors in production programme introduced by larger imperfections in consumption prediction and sources outages – Fig. 3.

This regulation affects the change of active power of generators in the whole range up to their withdrawal or connection into operation. It reacts to overall state of given power system and acts after the secondary power regulation or cannot be activated at all. The sources covering tertiary regulation of active power can use their whole regulation range or its parts for it. When starting from the zero power they have to supply power to the electric system equivalent to the basic regulation range of tertiary regulation (technical minimum of source). Reserve of tertiary regulation of active power can be secured with different activation times.

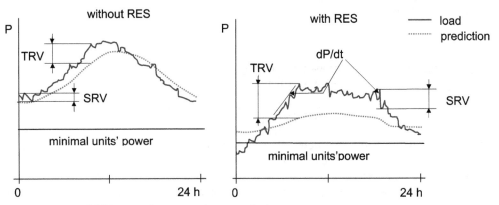

Fig. 3. Impact of RES operation on tertiary regulation reserve

Tertiary regulation *TRV30MIN* provides coverage of load changes caused by temperature, uncertainty in load estimates, outages of sources and electricity demand.

Necessary power reserve for tertiary power regulation coverage *TRV30MIN* can vary for both regulation directions and thus is split to positive and negative reserve.

Positive tertiary power regulation *TRV30MIN+* is calculated according to [1] and consists of several components:

Inaccuracy of load estimation and influence of temperature

$$TRV_{no} = NP_\phi \cdot MAX / 100 \tag{6}$$

(NP_ϕ – inaccuracy of load estimation, MAX – maximum load).
Stochastic load change:

$$TRV_{nz} = NP_{\phi+} \cdot MAX / 100 , \tag{7}$$

($NP_{\phi+}$ – positive inaccuracy of load estimation).
Substitution of tertiary power regulation in case of power production facility outage:

$$TRV_{vypbl} = SRV_{RGCE} . \tag{8}$$

Adjustment of the electricity market influence (this component can append value *TRV30MIN+* based on the historical data or expected changes depending on the electricity market).

The final value of *TRV30MIN+* is then calculated:

$$TRV30MIN+ = \sqrt{(TRV_{vypbl})^2 + (TRV_{nz})^2 + (TRV_{no})^2} \tag{9}$$

Negative tertiary power regulation *TRV30MIN-* is calculated according to [1] and consists of several components:

Inaccuracy of load estimation and influence of temperature

$$TRV_{no} = NP_\phi \cdot MAX / 100 \tag{10}$$

(NP_ϕ – inaccuracy of load estimation, MAX – maximum load).
Stochastic load change:

$$TRV_{nz-} = NP_{\phi-} \cdot MAX / 100 \tag{11}$$

($NP_{\phi-}$ – negative inaccuracy of load estimation).
The final value of $TRV30MIN\text{-}$ is then calculated:

$$TRV30MIN\text{-} = \sqrt{(TRV_{nz-})^2 + (TRV_{no})^2} \tag{12}$$

In view of some unpredictable fluctuations of RES power that can occur practically in the whole range of installed capacity, it is necessary to have sufficient regulation power of tertiary power regulation available at any time. Currently there are no statistical data for SPS which could be used to set starting values of increased tertiary regulation reserve caused by RES operation. That is why it would be suitable to use meteorological data as one of the supporting inputs for RES electricity production prediction.
Value of 30 minutes tertiary regulation reserve for RES considering a given degree of accuracy of RES production prediction is as follows:

$$TRV_{RES}^{30\,min\pm} = k_{NP} \cdot P_{inst\,RES} \, , \tag{13}$$

where $TRV_{RES}^{30\,min\pm}$ — increased 30 minutes tertiary regulation reserve caused by RES operation.
When calculating final values of $TRV30MIN$ it is necessary to distinguish winter and summer and also $TRV30MIN+$ or $TRV30MIN\text{-}$ services.

$$TRV_{Fin}^{30\,min+} = TRV_{RES}^{30\,min\pm} + TRV30MIN+ \tag{14}$$

$$TRV_{Fin}^{30\,min-} = TRV_{RES}^{30\,min\pm} + TRV30MIN- \tag{15}$$

Final values of tertiary regulation reserve TRV_{Fin} for different installed RES capacities and loads are in Table 4.

L_{max} (given) (MW)	1700						
L_{max} (calculated) (MW)	2261						
$P_{inst\,RES}$ (MW)	300	400	500	600	800	1000	1200
TRV_{no} (MW)	85	85	85	85	85	85	85
TRV_{nz+} (MW)	82	82	82	82	82	82	82
TRV_{vypbl} (MW)	62	62	62	62	62	62	62
$TRV30MIN+$ (MW)	133	133	133	133	133	133	133
$TRV_{RES}^{30\,min\pm}$ (MW)	84	112	140	168	224	280	336
$TRV_{Fin}^{30\,min+}$ (MW)	217	245	273	301	357	413	469

L_{max} (given) (MW)	1800						
L_{max} (calculated) (MW)	2394						
$P_{inst\,RES}$ (MW)	300	400	500	600	800	1000	1200
TRV_{no} (MW)	90	90	90	90	90	90	90
TRV_{nz+} (MW)	86	86	86	86	86	86	86
TRV_{vypbl} (MW)	66	66	66	66	66	66	66
$TRV30MIN+$ (MW)	141	141	141	141	141	141	141
$TRV_{RES}^{30min\pm}$ (MW)	84	112	140	168	224	280	336
TRV_{Fin}^{30min+} (MW)	225	253	281	309	365	421	477

L_{max} (given) (MW)	2400						
L_{max} (calculated) (MW)	3192						
$P_{inst\,RES}$	$SRV +/-$	$TRV30MIN-$ for different values of installed RES capacity					
(MW)	(MW)	10 %	20 %	40 %	60 %	70 %	80 %	90 %
300	140	148	157	173	190	199	207	215
400	145	151	162	185	207	218	229	241
500	152	154	168	196	224	238	252	266
600	158	157	173	207	241	257	274	291
800	171	162	185	229	274	297	319	341
1000	185	168	196	252	308	336	364	392
1200	198	173	207	274	341	375	409	442

L_{max} (given) (MW)	2600						
L_{max} (calculated) (MW)	3458						
$P_{inst\,RES}$	$SRV +/-$	$TRV30MIN-$ for different values of installed RES capacity					
(MW)	(MW)	10 %	20 %	40 %	60 %	70 %	80 %	90 %
300	145	160	168	185	202	210	219	227
400	151	163	174	196	219	230	241	252
500	157	165	179	207	235	249	263	277
600	163	168	185	219	252	269	286	303
800	177	174	196	241	286	308	331	353
1000	191	179	207	263	319	347	375	403
1200	204	185	219	286	353	387	420	454

Table 4. Final values of positive 30 minutes tertiary regulation reserve for different installed RES capacities and maximum loads L_{max}

4. Provision of auxiliary services and balancing electricity purchasing

Slovakia, as one of the members of interconnected European power system ENTSO-E, has to meet basic requirements for parallel operation of power systems. One of these basic requirements is also range and quality of auxiliary services, taking to account on one side global view of secure and reliable operation and on the other side, local nature of consumption of individual country. While the range of some service is strictly ordered from ENTSO-E (i.e. range of the primary regulation, which principally performs locally, reacts according to instantaneous frequency deviation, in tens of seconds, but impulse, i.e. origin of imbalance between electricity production and consumption can occur anywhere in interconnected system), ranges of other services can vary based on local behaviour of system and thus it has to meet only certain frame requirements.

Basically every power system must have power reserve secured for coverage of an outage of the largest source in order to be balanced in power in relation to other countries. This condition is secured by so-called stacking of various kinds of auxiliary services, or bi- or multi-lateral contracts with international partners, usually neighbouring countries.

In Slovak environment TSO SEPS acts as a partner for ENTSO-E. SEPS after consideration of before mentioned criteria elaborates i.a. a proposal of range of individual auxiliary services and in sense of valid legislation (as a regulated subject) submits to the national Regulatory Office for Network Industries, who in the form of decision defines range and price for a instantaneous availability of auxiliary services and balancing electricity

Fluctuation of electricity production from RES primarily imposes higher demands on regulation sources of SPS either on amount available reserves or on the quality of regulation. In the frame of interconnected system ENTSO-E the SPS is regulated for an agreed balance. SEPS has to keep the agreed value and quality of this regulation is monitored at the coordination centre level. Potential deterioration or violation of accepted standards in area of agreed balance would result in investigation of SPS by ENTSO-E.

4.1 Determination of necessary financial volume

In determination of necessary financial volume actual volumes for provision of individual kinds of auxiliary services are taken in account. These are in Table 5.

These values can be considered as maximums. In real operation they are not always achieved caused by various forces such as equipment failure, regular maintenance, financially underrated services (from the providers' point of view), unavailability on the market and other. Wider offer and competition on the auxiliary services market can be expected when new energy sources are being put into operation, e.g. combined cycle power plant Malženice, nuclear plant EMO 3 and 4 with expected favourable regulatory features and range.

Decision No. 0013/2010/E of the Regulatory Office for Network Industries determined prices and tariffs for auxiliary services provision for the time period of January 1, 2010 to December 31, 2010. Maximum prices for provision of auxiliary services are in Table 6.

Month	PRV±	SRV±	TRV30MIN+	TRV30MIN−	TRV3MIN+	TRV3MIN−	TRV120MIN
January	30	129	280	155	220	130	80
February	30	129	280	155	220	130	80
March	30	129	270	155	220	130	80
April	30	124	260	135	220	130	80
May	30	118	250	130	220	130	80
June	30	118	240	125	220	130	80
July	30	118	240	125	220	130	80
August	30	116	250	125	220	130	80
September	30	119	260	135	220	130	80
October	30	124	260	145	220	130	80
November	30	128	270	155	220	130	80
December	30	128	280	160	220	130	80

Table 5. Informative ranges of auxiliary services values (MW) for 2010

Auxiliary service	Price in € per MWh	Average range of auxiliary service (MW)
PRV	73,02	30
SRV	63,06	120
TRV3MIN+	17,59	220
TRV3MIN−	5,31	130
TRV30MIN+	16,92	250
TRV30MIN−	8,29	130
TRV120MIN	10,95	80

Table 6. Maximum prices for provision of individual auxiliary services

Deriving from actual range (Table 5.) and prices (Table 6.) of individual auxiliary services financial volume necessary for provision of reserved power in 2010 can be determined – so called payment for instantaneous availability according to the following formulae:

$$DE = O_{PpS} \cdot t_r \qquad\qquad [MWh, MW, h] \qquad (16)$$

$$RN = DE \cdot C \qquad\qquad [€, MWh, €/MWh] \qquad (17)$$

where DE is instantaneously available electric energy
O_{PpS} average range of auxiliary service
t_r number of hours per year
RN yearly costs
C price for auxiliary services provision

Calculated financial volume for PRV, SRV and TRV as well as costs to secure voltage regulation and auxiliary service "Black start" are in Table 7. Summing up yearly costs for instantaneous availability of individual auxiliary services shown in Table 7 overall yearly cost can be obtained for 2010, which is 183 594 016 €.

Auxiliary service	Yearly costs (€)
PRV	19 189 656
SRV	66 288 672
TRV3MIN+	33 899 448
TRV3MIN-	6 047 028
TRV30MIN+	37 054 800
TRV30MIN-	9 440 652
TRV120MIN	7 673 760
Voltage regulation	3 000 000
Black start	1 000 000
Sum	183 594 016

Table 7. Yearly cost of individual auxiliary services

Different scenarios of installed RES capacities rise (300, 400, 500, 600, 800, 1000 and 1200 MW) include also pressure on auxiliary services primarily to *SRV* and *TRV±*. The determination of accurate values with direct financial quantification is not simple as a number of unknown quantities are in play. To avoid placing a grave financial burden on consumers and excessively jeopardising power system operation appropriate effort will have to be given to harmonisation of these influences with the volume of *SRV* and *TRV±* with regard to actual increase of RES energy production and continuously with verified impact on power system.

Installed RES capacity (MW)	Rise of *TRV30MIN±* at per-cent supply from installed capacity to power system						
	10 %	20 %	40 %	60 %	70 %	80 %	90 %
300	8,4	16,8	33,6	50,4	58,8	67,2	75,6
400	11,2	22,4	44,8	67,2	78,4	89,6	100,8
500	14	28	56	84	98	112	126
600	16,8	33,6	67,2	100,8	117,6	134,4	151,2
800	22,4	44,8	89,6	134,4	156,8	179,2	201,6
1000	28	56	112	168	196	224	252
1200	33,6	67,2	134,4	201,6	235,2	268,8	302,4

Table 8. Increase of demand rise of *TRV30MIN±* for different scenarios of installed RES capacities increase

L_{max} **(given) (MW)**	**2200**						
L_{max} **(calculated) (MW)**	**2926**						
$P_{inst\,RES}$ **(MW)**	300	400	500	600	800	1000	1200
SRV_{UCTE} **(MW)**	77,5	77,5	77,5	77,5	77,5	77,5	77,5
$SRV_{DYN,L}$ **(MW)**	36,4	36,4	36,4	36,4	36,4	36,4	36,4
$SRV_{DYN,RES}$ **(MW)**	43	50	58	65	80	95	109
SRV_{FIN} **(MW)±**	133,8	139,3	146,0	152,0	165,4	179,2	192,4

Table 9. Final values of SRV at various scenarios of installed RES capacity increase and for L_{max} (given) 2200 MW

Increase of demand rise of $TRV30MIN\pm$ for different scenarios of installed RES capacities increase for different per-cent supplies from installed capacity is in Table 8. The amount of per-cent power supply to power system is influenced mainly by weather factor, which is due to global climate change becoming more and more unpredictable, for example May and June 2010 with having the most rainy days in the whole recorded period of weather observation in Slovakia (approximately 130 years). Cloudy weather without solar flux does not allow electricity supply from PV plants to power systems with any available installed capacity. Wind power plants have the advantage of not being directly dependent on solar flux and can produce electric energy during the whole day depending on wind conditions.

During summer season electricity produced by PV plants can be ideally supplied from 6:00 to 18:00, while in winter season from 9:00 to 15:00 with characteristic curve, where again these assumptions are subject to almost full solar flux. Without long term observations or long term acquired data from the operation these values can hardly be estimated.

For the necessary volume of auxiliary services it is also important whether sunny conditions last for a longer period of time or are unpredictably alternating with cloudy conditions.

In a longer period of sunny weather electricity supply from PV plants settles in daily cycles allowing distribution system operator to credibly implement this supply into DLD. Thus demand for range or activation of auxiliary services decreases. The opposite situation occurs in unstable weather with sunny spells. In this case the demand for range or activation of auxiliary services depending on installed or available power from RES will be enormous. From this point of view for the higher values of installed capacity in these sources bigger emphasis must be put on the possibility of operative increase of range of required auxiliary service, or make provision of non-guaranteed balancing electricity in exposed periods of time more flexible.

Installation of RES will require apart from increased volume of necessary regulation reserves also changes in actual system of procurement of auxiliary services by SEPS. Currently the substantial part of auxiliary service is procured in the frame of yearly selection procedure what appears as very ineffective for these kinds of sources. SEPS will be forced to procure large amounts of auxiliary services only at a frame of daily procurement or during the day as an auction of non-guaranteed balancing electricity. In this way contracted volumes of auxiliary services can be optimised while preserving or even enhancing operational security of power systems.

For determination of necessary financial volume for provision of auxiliary services at various scenarios of installed RES capacity increasing of $TRV\pm$ is considered from Table 8 and final values of SRV at various scenarios of installed RES capacity increase and for L_{max} (given) 2200 MW from Table 9. A modelling situation is considered, where for TRV± increase (Table 8) an average supply of 40 % from installed RES capacity during whole day is estimated. Furthermore, actually prices for auxiliary service provision stated by Regulatory Office for Network Industries and values ranges are used from Table 6.

After summing-up these considerations, sum of financial costs necessary for auxiliary services provision, considering RES putting in operation for various scenarios of installed capacities at the day of L_{max} (given) = 2 200 MW can be calculated.

Graphical comparison of the costs for auxiliary services provision at the day of L_{max} (given) = 2 200 MW for various scenarios of installed RES capacities against the costs for auxiliary services without RES are in Fig. 4.

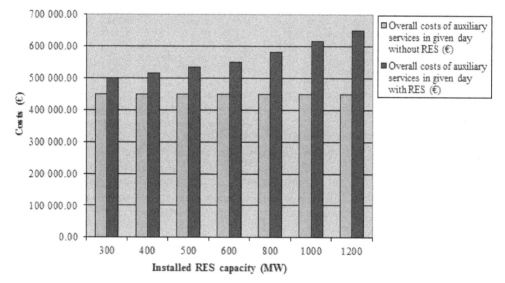

Fig. 4. Comparison of costs for instantaneous availability of auxiliary services at the day of L_{max} (given) = 2 200 MW for various scenarios of installed RES capacities and 40 % of supply

These values should be considered as the first estimation, without having possibility for results precising according to history. A simplification was used in calculations – use of averaged power supply from installed RES capacity during given day – in the amount of 40 % of average power supply. From this value a necessary range of $TRV30MIN\pm$ is derived. Based on at least one year data (better on couple of years data) acquired from RES in operation, this parameter can be precised, what will lead to more reliable estimations. Costs for auxiliary services are determined for the day of summer maximum for 24 hours. The reason is, that individual volumes of auxiliary services without RES are changing during the year and are depending on the load (expected maximum load was not mentioned in study entry values). Supplementary costs for auxiliary services for RES regulation are calculated only from estimated installed capacity and estimated production of RES.

Mentioned calculations shown that influence of RES upon the SEPS economics will be circa 10 mil. € per year, even at the lowest scenario.

Costs for auxiliary services are not the only costs that can be expected from RES installations caused by electricity production fluctuations. The highest costs will definitely be imposed to distribution systems operators. They will be charged for caused balance. Increasing of value of costs for balance in the whole SPS is hardly forecasted in advance, any calculations would be distorted. SEPS can expect also increase of additional costs for system operation in case of overloading of some parts of system due to energy production from RES, i. e. circular power flows, necessary network topology changes, re-dispatching of energy production (within the SPS or in adjacent regions) and other corrective measures. Their price can only be determined ex-post.

Graphical presentations of increase of costs for auxiliary services provision at various scenarios of installed RES capacities and for different per-cent supplies from RES capacities are depicted in Figs. 5 and 6. Values of auxiliary services, $TRV30MIN$ and SRV are in Tables 8 and 9. For costs calculation an actual prices of auxiliary services are used from Table 6. Yearly costs for instantaneous availability of auxiliary services forced by RES are on Figs. 7 and 8.

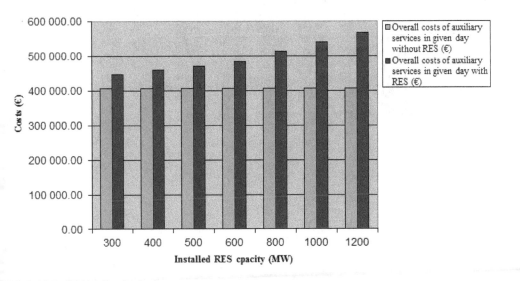

Fig. 5. Comparison of costs for auxiliary services at the day of L_{max} = 1 700 MW for various scenarios of installed RES capacities and 20 % of supply

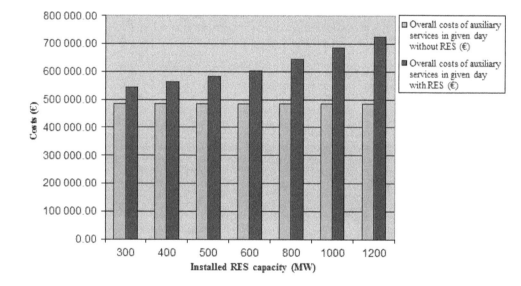

Fig. 6. Comparison of costs for instantaneous availability of auxiliary services at the day of L_{max} = 2 600 MW for various scenarios of installed RES capacities and 60 % of supply

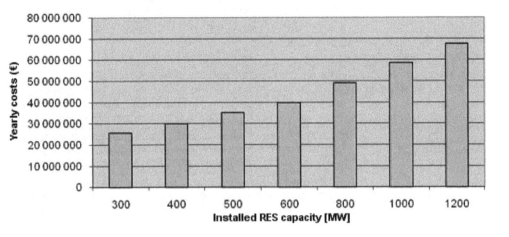

Fig. 7. Yearly costs for instantaneous availability of auxiliary services forced by RES at 10 % of supply from installed capacity

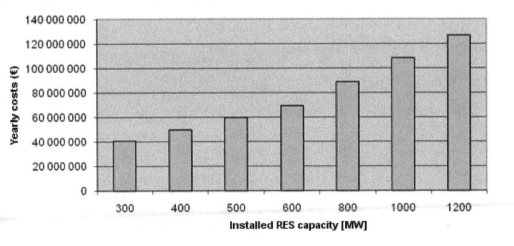

Fig. 8. Yearly costs for instantaneous availability of auxiliary services forced by RES at 90 % of supply from installed capacity

5. Results of analyses

Taking into account before mentioned analysis, the calculations were performed to demonstrate the influence of 300 – 1 200 MW RES connection on transmission system. These calculations were accomplished for three scenarios with different spectrum of sources. Results can be briefly summarized as follows:

Scenario A – year 2010. From the point of view of DLD covering by own sources some power shortages occurs, mainly for combinations at higher absolute minimums, i. e. 2 400 MW and 2 600 MW and for lower power supply from RES. DLD is fully covered for higher values of power supply from RES. The situation is different from the point of view of auxiliary services covering. Possible operation of RES is restricted to 60 – 120 MW at higher system loads, when auxiliary services are reserved mainly for the outages of traditional sources (non RES).

Scenario B – year 2011. In this scenario a new source has been put in operation – combined cycle Malzenice, which provides power for DLD coverage and also a good deal of auxiliary services. The situation from the point of view of auxiliary services covering is partially improved. Theoretically, secure operation of RES is possible with powers of 250 – 300 MW.

Scenario C – year 2013. In this scenario, due to large additional installed capacity in the new sources, SPS will be fully self-sustaining in DLD covering and installed RES capacity can theoretically reach values up to 1 000 MW for any system load. Accomplished calculations have shown that any DLD can be covered and there are no problems with auxiliary services, but substantial problems will rise in electricity export. Specified technological limits for electricity export across border profiles can be found in SEPS web pages www.sepsas.sk. Maximum possible electricity export from Slovak transmission system is 2 000 MW. If this number is taking in account, i. e. considering maximum export of 2 000 MW, electricity production will be higher then allowed export in 90 % of calculated cases.

From the figures showing yearly costs for instantaneous availability of auxiliary services follows, that electricity price for end-consumers will depend on installed RES capacity, if actual electricity purchase prices and auxiliary services prices are valid.

6. References

[1] SEPS, a.s.: *Technical Rules Establishing Technical Design and Operational Requirements for Connection to the Transmission System of the Operator SEPS, a.s.*, http://www.sepsas.sk, July 2009

[2] SEPS, a.s.: *Operating Instructions of the Transmission System Operator SEPS, a.s.*, July 2011, http://www.sepsas.sk

[3] SK Act No. 656/2004 Coll. *on Energy and Consequential Amendments,*

[4] SK Act No. 309/2009 Coll. *on the Promotion of Renewable Energy Sources and High-Efficiency Cogeneration and on Amendments to Certain Acts*

[5] SK Act No. 276/2001 Coll. *on Regulation in Network Industries and on Amendments and Additions to Some Acts*

[6] SK Government Decree No. 317/2007 Coll. *on the Regulation of the Electricity*

[7]   European Community Regulation 1228/2003/EC of the European Parliament and of the Council of 26 June 2003 on *Conditions for Access to the Network for Cross-Border Exchanges in Electricity*

[8]   ENTSO-E RG CE, *Operation Handbook*, June 2008, http://www.entsoe.eu, 2010

[9]   Influence of Wind Power Plants on Slovak Power System, Žilina, June 2008

[10] Operational Instruction No. 433-3 *Voltage Regulation in Slovak Power System*, December 2009, SEPS a.s.

[11] *Decisions of Regulatory Office for Network Industries*, http://www.urso.gov.sk

[12] Griger, V.; Gramblička, M.; Novák, M.; Pokorný M.: *Operation, Control and Testing of Interconnected Power System*, EDIS, Žilina, 2001

[13] Šmidovič, R., Rapšík, M., Novák, M.: *Auxiliary Services and Balancing Electricity*, EE časopis pre elektrotechniku, Bratislava, 6/2007, pp.: 5 – 8, ISSN 1335-2547

[14] *Rules for the Transmission System Operation Extract from the Grid Code*, Revision 09, ČEPS a.s., (January 2009)

[15] Novák, O., Strnad, T., Horáček, P., Fantík, J.: *Planning of Ancillary Services Securing Power System*, Proceedings of IEEE Conference on Environment and Electrical Engineering. Karpacz, Poland, 2009

[16] Burton, T., Sharpe, D., Jenkins, N., Bossanyi, E.: *Wind Energy Handbook. John Wiley & Sons*, Ltd, 2001

[17] *Reliability Standards for the Bulk Electric Systems of North America*, North American Electric Reliability Corporation, Princeton, NJ (2008)

[18] Havel, P., Horáček, P., Černý, P., Fantík, J.: *Optimal Planning of Ancillary Services for Reliable Power Balance Control*, IEEE Trans. on Power Systems, Vol. 23, No. 3, 1375-1382 (2008)

[19] El-Tamaly, H.,H., El-Baset Mohammed, A.: *Impact of Interconnection Photovoltaic/Wind System With Utility on Their Reliability Using a Fuzzy Scheme*, The 3rd Minia International Conference for Advanced Trends in Engineering, El-Minia, Egypt, 3-5 April, 2005

[20] Tan, Y., T., Kirschen, D., S., Jenkins, N.: *Impact of a Large Penetration of Photovoltaic Generation on the Power System*, 17th International Conference on Electricity Distribution, Barcelona, May 2003

[21] Kilk, K., Valdma, M.: *Determination of Optimal Operating Reserves in Power Systems*, Oil Shale, Vol. 26, No. 3 Special, pp.220-227, 2009

[22] *Damas Energy*, Information System of SEPS, a.s., https://dae.sepsas.sk/, 2010

[23] Online Monitoring of Photovoltaic Power Plants, http://www.sollaris-sk.sk/, 2010

[24] *Online Monitoring of Photovoltaic Power Plants*, http://www.htmas.eu/, 2010

[25] Heinemann, D., Lorenz, E., Girodo, M.: *Solar Irradiance Forecasting for the Management of Solar Energy Systems*, Energy and Semiconductor Research Laboratory, Energy Meteorology Group, Oldenburg University

[26] Lexmann, E.: *Meaning of Used Formulations in Weather Predictions and Success Rate of Predictions*, SHMÚ Bratislava, www.shmu.sk, 8.1.2010

[27] http://www.shmu.sk, 4. 4. 2008.

[28] Landberg, L. et al.: *Short-term Prediction of Regional Wind Power Prediction*, Final report for The European Commission in the framework of the Non Nuclear Energy Programme JOULE III, December 1999, Contract JOR3-CT97-0272 PL971254.

[29] Petersen, E.L., Mortensen, N.G., Landberg, L., Hojstrup, J.H., Frank, H.P.: *Wind Power Meteorology*, Riso National Laboratory, Roskilde, Denmark, December 1997.

[30] http://www.wasp.dk/, 2. 6. 2008.
[31] STN EN 61400-12 (333160): 2001. *Wind Power Plants. Part 12: Power of Wind Power Plants Measurements.*

7

Highly Efficient Biomass Utilization with Solid Oxide Fuel Cell Technology

Yusuke Shiratori[1,2], Tran Tuyen Quang[1], Yutaro Takahashi[1],
Shunsuke Taniguchi[1,3] and Kazunari Sasaki[1,2,3]
[1]*Department of Mechanical Engineering, Faculty of Engineering, Kyushu University*
[2]*International Institute for Carbon-Neutral Energy Research (WPI), Kyushu University*
[3]*International Research Center for Hydrogen Energy, Kyushu University*
Japan

1. Introduction

Mankind has been consuming plants, i.e. biomass, as an energy source for living and developing on earth from the paleolithic period to early the modern period. Consumption of bio-energy does not change the atmospheric environment because carbon dioxide emitted by the use of bio-energies will be used by plants through the photosynthesis (Züttel, 2008). Since 1769 James Watt significantly improved the steam engine, invented by Thomas Savery in 1698. The steam engine was widely introduced for producing mechanical work from chemical energy of fuels, i.e. mineral coal and wood. More practical heat engines, external and internal combustion engines, have served for developing of human society for almost two and a half centuries. Since the Otto-Langen engine was first introduced in 1867, human society has developed using the internal combustion engines (IC engines), which nowadays are used worldwide for transportation, manufacture, power generation, construction and farming. However, large consumption of fossil fuels may bring about environmental pollution and climate change.

Fuel cells are electrochemical devices that convert chemical energy of fuels directly into electrical energy without the Carnot limitation that limit IC engines. Even in the smallest power range of less than 10 kW, fuel cells exhibit electrical efficiencies of 35-50 %LHV (lower heating value), while being silent, whereas engines and microturbines show low electrical efficiency of 25-30%LHV and high levels of noise. Therefore, the fuel cell which can be operated with very low environmental emission levels, is regarded as a promising candidate for a distributed power source in the next generation. Although most fuel cells operate with hydrogen as a fuel, solid oxide fuel cells (SOFCs) operated in a high temperature range between 600 and 900 ºC accept the direct use of hydrocarbon fuels. Hydrocarbon fuels directly supplied to SOFCs are reformed in the porous anode materials producing H_2-rich syngas, which is subsequently used to generate electricity and heat through electrochemical oxidation (Steele & Heinzel, 2001; Sasaki & Teraoka, 2003). This type of SOFC, so called internal reforming SOFC (IRSOFC), enables us to simplify the SOFC system. Electrochemical performances of IRSOFC have been reported for gaseous and liquid fossil fuels such as methane (Park et al., 1999), propane (Iida et al., 2007), n-dodecane

(Kishimoto et al., 2007), synthetic diesel (Kim et al., 2001), crude oil and jet fuel (Zhou et al., 2004). Highly efficient fuel cells operated by fossil fuels can certainly contribute to the suppression of environmentally harmful emissions, but in view of exhaustion of fossil resources, the utilization of renewable bio-energies should be more promoted. Direct feeding of biofuels to SOFC gives an environmental-friendly, compact and cost-effective energy conversion system. Biogas derived primarily from garbage is one of the most attractive bio-energies for SOFC (Van herle et al., 2004a; Shiratori et al., 2008, 2010a, 2010b). Recently, Shiratori et al. (2010) has demonstrated the stable operation of an IRSOFC operating on non-synthetic biogas over one month using an anode-supported button cell. On the other hand, the use of liquid biofuels is also attractive due to their easy storage and transportation with high energy density. Tran et al. (2011) has demonstrated the stable operation of an IRSOFC operating on practical palm-biodiesel over 800 h, also using an anode-supported button cell.

In this chapter, performances of IRSOFCs operating on biofuels are summarized and roadblocks to overcome for the realization of this type of highly-efficient carbon-neutral fuel cell are mentioned.

2. Sustainable society using internal reforming SOFC (IRSOFC) running on biofuels (*Bio*-SOFC)

Although the conventional large scale power system provides us with a stable electric power supply, the associated large consumption of fossil fuels and release of large amount of waste heat are unfit for social and environmental needs in recent years. Now, the role of biomass, having the largest exploitation potential among renewable energy resources, becomes very important. Of course, the use of edible plants is highly restricted, but the use of organic wastes is highly desirable. Biogas and biodiesel fuels (BDFs) are attractive alternative fuels which can be produced from bio-wastes, and their spread will generate synergistic effects to create new industries and employment in their production and refinement processes.

Among fuel cells, the SOFC is the only technology capable of converting bio-energies directly to electricity without an external fuel reformer (Staniforth et al. 1998). As for the low temperature fuel cells like polymer electrolyte fuel cell (PEFC), the external reforming process is essential prior to electrochemical conversion of biofuels to electricity. Superiority of an IRSOFC running on biofuels (hereafter called *Bio*-SOFC) is described in Fig. 1. By selecting red arrows in this figure social needs are satisfied. Our final goal is to establish a microgrid system as shown in Fig 2 using *Bio*-SOFC as major distributed generators providing heat and power on site. These distributed *Bio*-SOFCs can contribute to leveling of the unstable power supply from solar and wind energies. However, breakthroughs are necessary to realize the *Bio*-SOFC system.

3. Solid oxide fuel cell (SOFC)

The solid oxide fuel cell (SOFC) offers a highly efficient and fuel-flexible technology for distributed power generation and combined heat and power (CHP) systems, and it is obviously promising technology for utilizing biofuels. In this section, SOFC technologies are briefly reviewed from fundamentals to current status of development.

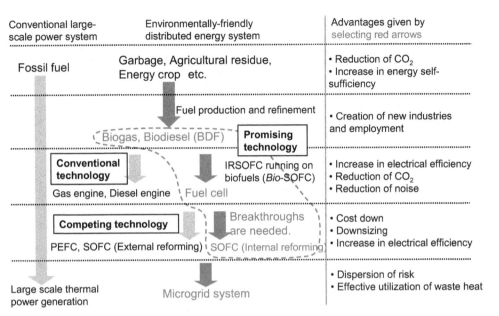

Fig. 1. Superiority of IRSOFC running on biofuels (*Bio*-SOFC), a promising candidate as a distributed generator in the next generation.

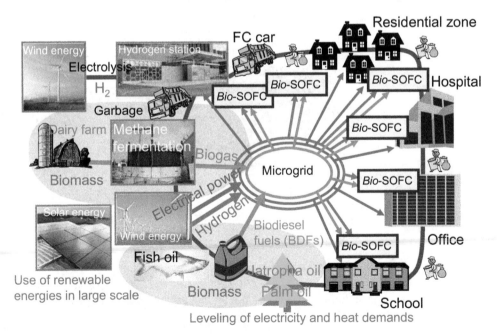

Fig. 2. Microgrid system using *Bio*-SOFC as a major distributed power source.

3.1 Operation mechanism

A fuel cell in general converts chemical energy of the fuel directly into electrical energy without converting it to mechanical energy. Therefore, the fuel cell has potential of attaining higher electrical conversion efficiency than those of conventional technologies such as heat engines limited by Carnot efficiency. Fig. 3 shows the principle of SOFC operation. The basic unit of an SOFC, i.e. cell, consists of an electrolyte sandwiched with two electrodes, anode and cathode. In the electricity generation process, an oxide ion O^{2-} is generated from oxygen in air via the cathodic reaction (1).

$$\tfrac{1}{2} O_2 \text{ (g)} + 2e^- \rightarrow O^{2-} \tag{1}$$

Normally, SOFCs are operated in the temperature range between 600 and 900 °C in which electrolytes composed of doped metal oxides can exhibit rather high oxygen ion conductivity. Oxygen ions generated in the cathode are transported to the anode side through the dense electrolyte and are used to electrochemically oxidize a fuel, here hydrogen, in the anodic reaction (2).

$$H_2 \text{ (g)} + O^{2-} \rightarrow H_2O \text{ (g)} + 2e^- \tag{2}$$

$$CO \text{ (g)} + O^{2-} \rightarrow CO_2 \text{ (g)} + 2e^- \tag{3}$$

In the high temperature SOFC, not only hydrogen but also carbon monoxide can contribute to the generation of electricity (3). Hydrogen and carbon monoxide can be produced by steam reforming or partial oxidation of hydrocarbon fuels on the Ni-based anode material, therefore in principle hydrocarbon fuels can be directly supplied to SOFC without using a pre-reformer.

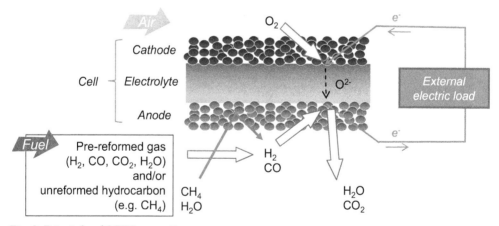

Fig. 3. Principle of SOFC operation.

The electromotive force of fuel cell, E, derived from the difference in the partial pressure of oxygen, $p(O_2)$, between cathode and anode sides can be expressed by the Nernst equation (4).

$$E = (RT/4F) \ln \{ p(O_2, \text{ c})/p(O_2, \text{ a})\}, \tag{4}$$

where R is the gas constant and T is absolute temperature. a and c denote anode and cathode sides, respectively. Theoretically, E is approximately 1 volt, and the ideal electrical efficiency can be calculated by $\Delta G/\Delta H$, where ΔG and ΔH are Gibbs free energy change and enthalpy change of the hydrogen combustion reaction (5), respectively. At the operating temperature of 800 °C ideal efficiency becomes 70%LHV.

$$H_2 (g) + \frac{1}{2} O_2 (g) = H_2O (g) \tag{5}$$

The actual electrical efficiency of an SOFC (40-50 %LHV) is always lower than the ideal value because fuel utilization (U_f) can not be increased up to 100 % in the practical SOFC system and the contribution of the internal resistances such as resistances of the materials themselves, contact resistances and electrode reaction resistances is not negligible. However, in the small size fuel cell systems, the heat generation, including the intrinsic heat release, ΔH -ΔG, can be utilized effectively on site leading to overall efficiency above 80 %LHV.

The SOFC has the following advantages because of its high operating temperature. Various kinds of fuel, such as natural gas, liquefied petroleum gas, kerosene and biofuels, etc. can be utilized with a simple fuel processing system. Even direct feeding of such practical fuels is theoretically possible. Higher electrical efficiency above 50 %LHV can be obtained by setting higher fuel utilization. Overall efficiency can be enhanced by using the heat released from the cell for the fuel reforming process, in which endothermic steam reforming proceeds as a main reaction. This kind of energy recycle is possible because the operational temperature of SOFC is nearly the same as that of the reformer. In addition, further enhancement of electrical efficiency is expected by using residual fuel and water vapor in a downstream gas turbine and steam turbine. High quality heat from the high temperature SOFC system can also be utilized effectively for hot water supply as well as reformer.

3.2 Component materials of SOFC and stack configurations

In Table 1, requirements for component materials of a cell and interconnector are summarized. Typical materials are also listed in this table. The electrolyte has to be gas-tight to prevent leakage of fuel and oxidant gases. Both electrodes have to be porous to provide electrochemical reaction sites. The interconnector plays a role of electrically connecting the anode of one cell and the cathode of the adjoining cell, and also separating the fuel in the anode side from the oxidant in the cathode side. Component materials must be heat resistant and durable in the highly oxidative and reductive atmospheres for cathode and anode sides, respectively.

To fabricate a cell, powders of these component materials are formed into desired shapes by general ceramic processing such as extrusion, slip casting, pressing, tape casting, printing and dip coating (Stöver et al., 2003). Subsequently, the resulting "green" ceramics undergo heat-treatments. High temperature sintering processing above 1300 °C is normally required to obtain a dense and gas tight electrolyte layer.

In an SOFC, all solid state fuel cell, various types of cell configuration have been designed and classified by support materials and shapes as summarized in Table 2. The SOFC has a laminate structure of thin ceramic layers, therefore a support material is necessary to ensure the mechanical stability. Anode-supported, electrolyte-supported, cathode-supported, metal-supported and nonconductive ceramic-supported (segmented-in-series) types have been developed. From the viewpoint of cell shape, there are roughly three types, i.e., planar, flat tubular and tubular types.

Component materials		Requirements	Typical materials
Cell	Electrolyte	Dense (Gas tight), Ionically conductive	Y_2O_3-stabilized ZrO_2 (YSZ), Sc_2O_3-stabilized ZrO_2 (ScSZ), Gd_2O_3 doped CeO_2 (GDC), $(La,Sr)(Ga,Mg)O_3$ (LSGM))
	Anode	Porous, Electrochemically active, Electronically conductive	Ni-YSZ, Ni-ScSZ, Ni-GDC
	Cathode	Porous, Electrochemically active, Electronically conductive	$(La,Sr)MnO_3$, $(La,Sr)(Fe,Co)O_3$
Interconnector		Dense (Gas tight), Electronically conductive	$(La,Sr)CrO_3$, $(La,Ca)CrO_3$ $(Sr,La)TiO_3$, Stainless steel

Table 1. Requirements for component materials and typical materials used to fabricate SOFC.

Support materials	Cell shape		
	Planar	Flat tubular	Tubular
Anode-supported	Versa Power Systems, Ceramic Fuel Cells, Topsoe Fuel Cell, Nippon Telegraph and Telephone, NGK Spark Plug	Kyocera	Acumentrics, TOTO
Electrolyte-supported	Hexis, Mitsubishi Materials		
Cathode-supported		Siemens Power Generation	Siemens Power Generation, TOTO
Metal-supported	Ceres Power		
Nonconductive ceramic-supported (Segmented-in - series *)		Rolls-Royce Fuel Cell Systems, Tokyo gas	Mitsubishi Heavy Industries

* Cells are formed in series on a nonconductive porous ceramics.

Table 2. Various SOFC configurations.

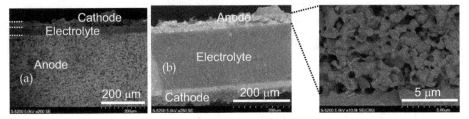

Fig. 4. FESEM images of (a) anode-supported cell and (b) electrolyte-supported cell.

Resistance of the electrolyte dominates the internal voltage loss in many cases, thus making thinner electrolytes is a key technology to achieve better performance especially at lower operating temperatures. For example, an anode-supported cell (Fig. 4a), in which a thin electrolyte layer with a thickness of around 15 μm is formed on the anode substrate with a thickness of 1 mm, enables operation of an SOFC around 200 K lower in temperature as compared to an electrolyte-supported cell (Fig. 4b) (Steele & Heinzel, 2001).

There are also various types of stack structures with different flow directions of electric current, fuel and air, as shown in Fig. 5. The planar type (Fig. 5a) can achieve higher current density and lower manufacturing cost, but it has lower tolerance to thermal stress. Flat tubular type (Fig. 5b) exhibits higher durability than planar type because the area for gas seal is smaller compared to planar type without changing the direction of electric current. This type enables downsizing of a SOFC stack but is not suitable for hundred kW class large systems due to the mechanical properties of the interconnector material. The segmented-in-series type (Fig. 5c) has the advantage of scalability because gas seals and electrical connections between the adjacent cells are already completed. However, this structure is complicated and a lot of optimization is required for precise fabrication.

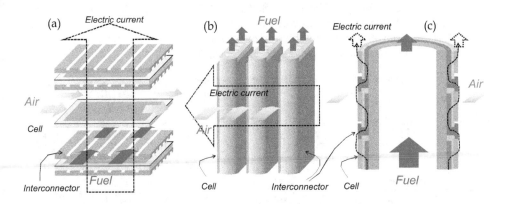

Fig. 5. Structures of SOFC stack; (a) anode-supported planar type (b) anode-supported flat tubular type and (c) segmented-in-series tubular type.

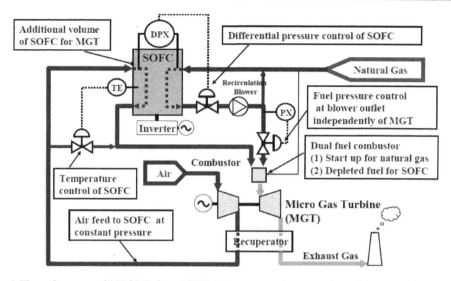

Fig. 6. Flow diagram of 200 kW-class SOFC-micro gas turbine combined system developed by Mitsubishi Heavy Industries (Yoshida et al., 2011).

Large scale SOFC systems are being developed aiming for a distributed electrical power plant with high energy conversion efficiency. Mitsubishi Heavy Industries (JP) has developed a 200 kW class micro gas turbine hybrid system as shown in Fig. 6 with a maximum efficiency of 52.1% LHV, and a maximum gross power of 229 kW-AC was achieved with natural gas as a fuel. Their final goal is to achieve electrical efficiency of 70% by developing large-scale power generation system in which the SOFC integrates with gas turbines and steam turbines (Yoshida et al., 2011). Rolls-Royce Fuel Cell Systems (GB) is designing a stationary 1 MW SOFC power generation system based on their segmented-in-series cell stack named Integrated-Planar SOFC technology (Haberman et al., 2011; Gardner et al., 2000). FuelCell Energy (US) is developing SOFC power plants, currently utilizing SOFC stacks developed by Versa Power Systems (CA). Their ultimate goal is to develop Multi-MW SOFC power plants suitable for integration with coal gasifiers and capable of capturing > 90% of carbon in coal syngas (Huang et al., 2011).

Fig. 7. Appearances of installation sites of residential SOFC CHP systems under demonstrative research project in Japan (Hosoi et al., 2011).

Small scale SOFCs of 1 -2 kW class are now being developed all over the world, aiming at early commercialization of residential CHP systems (Fig. 7). City gas is generally used as a fuel with high overall efficiency more than 80 %LHV obtained by utilizing both electricity and heat on site. In Japan, a demonstrative research project is now being carried out in which Kyocera (JP), Tokyo gas (JP) and TOTO (JP) participate as manufacturers of SOFC stacks. More than 200 systems have been installed on actual residential sites as of FY2010. In this project, reductions of primary energy consumption and CO_2 emission by 16 and 34 %, respectively, were demonstrated (Hosoi et al., 2011). Until now, long-term operation over 25,000 h has been demonstrated, and potential of 40,000 h durability has been confirmed. Ceramic Fuel Cells (AU) is manufacturing the residential system called BlueGen which can deliver initial electrical efficiency of 60 %LHV at 1.5 kW-AC, and exhibited 55 % efficiency after 1 year operation (Föger et al., 2010; Payne et al., 2011). Hexis Ltd. (CH) had operated 1 kW class system for 28,000 h (Mai et al., 2011). Acumentrics (US) (Byham et al., 2010) and Ceres Power (GB) are also developing residential SOFC systems (Leah et al., 2011). For the spread of these SOFC systems, further enhancements of electrical efficiency, fuel flexibility and thermomechanical reliability are essential. In this chapter, these challenges are summarized taking our effort, application of biofuels to SOFC, as a good example of advancements.

4. Performance of IRSOFC operating on biofuels (*Bio*-SOFCs)

4.1 Experimental
4.1.1 SOFC single cell used in the tests

Ni-yttria stabilized zirconia cermet (Ni-YSZ) is the most popular material for SOFC anodes. However, it has been reported that Ni-scandia stabilized zirconia (Ni-ScSZ) offers better catalytic performance and can suppress coking when hydrocarbon fuels are supplied to SOFC (Ke et al., 2006). Thus, we used anode-supported type cells based on the anode/electrolyte bi-layer (half cell) of Ni-ScSZ/ScSZ to evaluate the electrochemical and catalytic performance of IRSOFCs operating on biofuels (*Bio*-SOFCs). Two types of half cells, button cell with diameter of 20 mm and square-shaped cells with area of 5 x 5 cm² purchased from Japan fine ceramics were used to fabricate single cells. The cells consist of ScSZ electrolyte with a thickness of 14 μm sintered on a porous anode support (mixture of NiO and ScSZ (NiO:ScSZ = 5.6:4.4)) with a thickness of 800 μm. A mixture of $(La_{0.8}Sr_{0.2})_{0.98}MnO_3$ (> 99.9 %, Praxair, USA, abbreviated by LSM) and ScSZ (Daiichi Kigenso Kagaku Kogyo, Japan) with a weight ratio of 1:1 was adopted as a cathode functional layer and coarse LSM was applied as a cathode current collector layer. Porous cathodes with the area of 0.8 x 0.8 and 4 x 4 cm² for button and square-shaped cells, respectively, were obtained by sintering process at 1200 ºC for 3 h. Component materials of the single cells were summarized in Table 3.

Component materials	Composition	Thickness / μm
Electrolyte	10 mol% Sc_2O_3-1 mol% CeO_2-ZrO_2 (ScSZ)	14
Anode functional layer	56%NiO-44%ScSZ	26
Anode substrate	56%NiO-44%ScSZ	800
Cathode	50%$(La_{0.8}Sr_{0.2})_{0.98}MnO_3$ (LSM)-50%ScSZ	~30
Cathode current collector layer	LSM	~30

Table 3. Component materials of the anode-supported single cells (Shiratori et al., 2010b).

4.1.2 IRSOFC operating on biogas

The SOFC testing system and automatic gas chromatograph were connected to the methane fermentation reactor placed in Tosu Kankyo Kaihatsu Ltd. as shown in Fig. 8 (Shiratori et al., 2010a). Garbage collected in Tosu-city of Saga prefecture was mixed with water resulting in waste slurry. After materials unsuitable for anaerobic fermentation were filtered out, cattle manure was added to the slurry followed by the treatment with acid and methane fermentation processes to produce biogas (mixture of CH_4 and CO_2) containing 790 ppm H_2S. The raw biogas was passed through a desulfurizer packed with FeO pellets 20 cm³ in size. The typical composition of desulfurized biogas sampled from the fermentation reactor is listed in Table 4. The concentration of H_2S was less than 0.5 ppm. The concentrations of the other fuel impurities, CH_3SH, Cl_2, HCl, NH_3 and siloxane, were below the detection limits (2 ppb, 60 ppb, 0.4 ppm, 0.6 ppm and 10 ppb, respectively), indicating that this gas can be fed directly into a SOFC (Haga et al., 2008). In any case, 1 ppm level H_2S contamination must be taken into account even after desulfurization treatment. The experimental setup for testing IRSOFC operating on biogas has been described elsewhere (Shiratori et al., 2010a). The pressure controlled real biogas (0.1 MPa) was directly distributed to the SOFC at 800 °C and the gas chromatograph with flow rates of 25 and 140 ml min⁻¹, respectively, in order to evaluate the electrochemical performance with simultaneous monitoring of biogas composition. In this experiment, water vapor in the real biogas was removed by a cold trap thermostated at 0 °C. Dry air was supplied to the cathode side with a flow rate of 50 ml min⁻¹

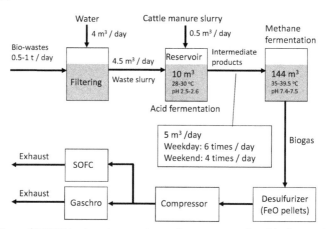

Fig. 8. Connection of SOFC test system and gas chromatograph with the methane fermentation reactor in Tosu-city, Japan (Shiratori et al., 2010a, 2010b).

Gaseous species	Concentration
CH_4	62.6 vol %
CO_2	35.7 vol %
H_2O	1.62 vol %
N_2	0.09 vol %
H_2	99 vol ppm
H_2S	< 0.5 ppm

Table 4. Typical composition of the actual desulfurixed biogas (Shiratori et al., 2010a).

4.1.3 IRSOFC operating on biodiesel fuels (BDFs)

Palm-, jatropha- and soybean-biodiesel fuels (BDFs) were produced from refined palm-, jatropha- and soybean-oils, respectively, at Bandung Institute of Technology, Indonesia (ITB). The main chemical components of the BDFs are listed in Table 5.

Formula	Structure	Concentration / wt %		
		Palm-BDF	Jatropha-BDF	Soybean-BDF
Myristic acid methyl ester $C_{15}H_{30}O_2$ (C14:0)		2.73	0.07	0.14
Palmitic acid methyl ester $C_{17}H_{34}O_2$ (C16:0)		39.1	14.6	11.0
Stearic acid methyl ester $C_{19}H_{38}O_2$ (C18:0)		4.21	6.55	3.21
Oleic acid methyl ester $C_{19}H_{36}O_2$ (C18:1)		40.4	39.9	24.4
Linoleic acid methyl ester $C_{19}H_{34}O_2$ (C18:2)		12.2	37.6	54.0
Linolenic acid methyl ester $C_{19}H_{32}O_2$ (C18:3)		0.26	0.19	5.75

Table 5. Main components in the biodiesel fuels applied to SOFC (Shiratori et al., 2011).

Table 6 shows the concentrations of saturated and unsaturated components in the BDFs and their average structures. Palm-BDF consisted mainly of 46.8 % saturated fatty acid methyl esters (FAMEs) and 40.8 % mono-unsaturated FAMEs. Jatropha- and soybean-BDFs contained a higher amount of unsaturated FAMEs compared to palm-BDF. Jatropha-BDF consisted mainly of 40.9 % mono-unsaturated FAMEs and 37.6 % di-unsaturated FAME. Soybean-BDF consisted mainly of 24.8 % mono-unsaturated FAMEs and 54.0 % di-unsaturated FAME. Soybean-BDF not only contained higher amount of di-unsaturated FAME compared to jatropha-BDF but also contained a rather high amount of tri-unsaturated FAME (5.75 % linolenic fatty acid methyl ester). All BDFs had almost the same physical properties similar to petro-diesel. The experimental setup for testing the IRSOFC running on BDFs has been described elsewhere (Shiratori et al., 2011; Tran et al., 2011). Dry air was supplied to the cathode side with the flow rare of 150 ml min⁻¹. BDFs and H_2O were supplied to the anode side with the flow rates of 6 and 21-22 µl min⁻¹, respectively. BDFs were evaporated and mixed with H_2O at 600 °C and then fed to the anode side together with nitrogen carrier gas with the flow rate of 50 ml min⁻¹.

	Concentration / wt %		
	BDFs		
	Palm-BDF	Jatropha-BDF	Soybean-BDF
Saturated	46.8	21.4	15.4
Unsaturated	53.2	78.6	84.6
Mono-	40.8	40.9	24.8
Di-	12.2	37.6	54.0
Tri-	0.26	0.19	5.75
Average structure	$C_{18.0}H_{34.8}O_2$	$C_{18.7}H_{35.0}O_2$	$C_{18.8}H_{34.5}O_2$

Table 6. Composition of saturated and unsaturated components in the tested BDFs (Shiratori et al., 2011).

The C-H-O diagram (see Fig. 9) clearly shows that unless a generous amount of oxidant is added, biogas (CH_4/CO_2 = 1.5) and BDF will form coke on the anode material during SOFC operation. We added air and water to gaseous biogas and liquid BDF, respectively, to avoid carbon deposition from the thermodynamic point of view.

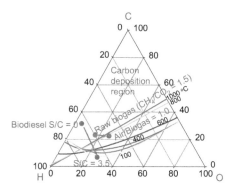

Fig. 9. C-H-O ternary diagram showing the possibility of coking when the biofuels are fed directly to SOFC (Sasaki & Teraoka, 2003).

4.2. Long term test of *Bio*-SOFCs
4.2.1 IRSOFC operating on biogas
The main reactions involved in an IRSOFC running on biogas are listed in Table 7. Internal dry reforming of CH_4 (reaction 1) proceeds on a Ni-based anode using CO_2 inherently included in biogas without an external reformer. Then, the produced H_2 and CO are electrochemically oxidized to produce electricity (reactions 2 and 3) (Shiratori et al., 2009a).

1	$CH_4 + CO_2 \rightarrow 2CO + 2H_2$	Dry reforming
2	$H_2 + O^{2-} \rightarrow H_2O + 2e^-$	Electrochemical oxidation
3	$CO + O^{2-} \rightarrow CO_2 + 2e^-$	Electrochemical oxidation

Table 7. Main reactions involved in IRSOFC running on biogas (Shiratori et al., 2009a).

Only a few reports have provided the performance of IRSOFCs operating on biogas (Staniforth et al., 2000; Shiratori et al., 2008; Girona et al., 2009; Lanzini & Leone, 2010a), because carbon formation thermodynamically can take place on the anode material. Staniforth et al. (2000) has reported the results of direct-feeding of landfill biogas (general CH_4-rich biogas). However, that was a short term experiment and not continuous feeding of as-produced real biogas. To avoid coking, pre-reforming of biogas is generally required (Van herle et al., 2004b). Recently, development of a new fermentation path producing H_2-rich biogas (Leone et al., 2010) and highly active catalysts to assist biofuel reforming (Xuan et al., 2009; Yentekakis et al., 2006; Zhou et al., 2007) have been reported. Heretofore, we have succeeded in stable operation of an IRSOFC running on actual biogas produced in waste treatment center using Ni-ScSZ cermet as an anode material without any support catalysts.

Figure 10a shows the cell voltages of IRSOFC operating on biogas measured at 200 mA cm^{-2}. Simulated biogas with the CH_4/CO_2 ratio of 1.5 led to a stable cell voltage above 0.8 V for 800 h. The degradation rate of only 0.4 %/1000 h proved that the biogas-fueled SOFC can be operated with an internal reforming mode. For the real biogas generated from the methane fermentation reactor, rather high voltage comparable to that obtained by simulated biogas was achieved for 1 month, although there is a voltage fluctuation. Monitoring of biogas composition simultaneously with the cell voltage (Fig. 10b) revealed that voltage fluctuation (a maximum of 50 mV level) appeared in synchronization with the fluctuation of CH_4/CO_2 ratio. An abrupt increase in CO_2 concentration induced temporarily decreased cell voltage. The CH_4/CO_2 ratio in the real biogas fluctuated between 1.4 and 1.7 which corresponds to CH_4 concentration range between 58 and 63 vol%. The biogas composition is influenced by many factors, for example, type of organic wastes, physical states and operational conditions of methane fermentation such as temperature and pH of the waste slurry.

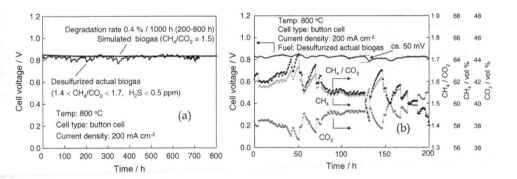

Fig. 10. Performance of IRSOFC operating on biogas (Shiratori et al., 2009a); (a) the results measured at 800 °C using anode-supported button cells and (b) voltage fluctuation in synchronization with the fluctuation of biogas composition during the long term test (initial 200 h of (a)).

Fig. 11. Anode-supported cells after long term test shown in Fig. 10a using (a) simulated and (b) practical biogases.

From the thermodynamic point of view, a CH_4/CO_2 ratio of 1.5 should cause coking at 800 °C, however carbon formation did not occur on the anode support for simulated biogas even after 800 h as shown in Fig. 11a. Provided that a fuel cell current is applied, this fuel composition (see Fig. 9) close to the border of the carbon deposition region may not cause coking. As for the real biogas, severe coking occurred during the long term test (Fig. 11b). The trigger of the coking was trace H_2S (< 0.5 ppm) in the real biogas which can cause deactivation of Ni catalyst for the dry reforming of methane (Sasaki et al., 2011; Shiratori et al., 2008). Acceleration of carbon deposition in the presence of trace H_2S was also reported for SOFCs operated with partially-reformed CH_4-based fuels (Yuki et al., 2009). Although a threshold of maximum H_2S concentration tolerance for Ni-stabilized zirconia cermet anode must be provided in future work, total sulfur concentration should be reduced to less than 0.1 ppm level by employing an optimum desulfurization process.

4.2.2 IRSOFC operating on biodiesel fuel (BDF)

The main reactions inferred in the steam reforming of BDF are listed in Table 8 (Nahar, 2010). Steam reforming (reaction 1) produces H_2 and CO, and pyrolysis (reaction 2) produces light hydrocarbons (C_xH_y) and coke as well as H_2 and CO may occur as competing reactions. Both reactions are endothermic, more promoted at higher temperatures. Steam reforming is a heterogeneous reaction catalyzed by Ni, whereas pyrolysis, a non-catalytic gas phase reaction, tends to occur at water-lean regions or on the deactivated surface of the Ni-based anode. Excess H_2O reacts not only with the light hydrocarbons (reaction 3) but also with the produced CO (reaction 4) to form further H_2. Reactions 5 and 6 are exothermic hydrogenation reactions which consume H_2 and produce CH_4. Reactions 7 and 8 are endothermic gasification reactions of coke. While S/C = 3.5 is thermodynamically out of the carbon deposition region (see Fig. 9), contributions of reactions 5, 7 and 8 are not negligible once carbon is deposited on the anode surface.

1	$C_nH_mO_2 + (n-2)H_2O \Leftrightarrow (n+m/2-2)H_2 + nCO$	Steam reforming
2	$C_nH_mO_2 \Leftrightarrow$ gases (H_2, CO, C_xH_y) + coke	Pyrolysis
3	$C_xH_y + xH_2O \Leftrightarrow xCO + (x+y/2)H_2$	Steam reforming
4	$CO + H_2O \Leftrightarrow H_2 + CO_2$	Water-gas shift
5	$C + 2H_2 \Leftrightarrow CH_4$	Hydrogenation
6	$CO + 3H_2 \Leftrightarrow CH_4 + H_2O$	Hydrogenation
7	$C + H_2O \Leftrightarrow CO + H_2$	Coke gasification
8	$C + CO_2 \Leftrightarrow 2CO$	Boudouard-reaction

Table 8. Main reactions involved in steam reforming of biodiesel fuel (Nahar, 2010).

Figure 12 shows the cell voltage of an IRSOFC operating on wet palm-BDF (S/C = 3.5) measured at 0.2 A cm^{-2}. Stable operation with a degradation rate of about 0.1 mV h^{-1} (approx. 1.5 %/100 h) was recorded. Total ohmic resistance of the cell, R_{IR}, total polarization resistance R_P and the internal resistance of the cell, R_{int} (the sum of R_{IR} and R_P), measured under open circuit condition are also plotted in Fig. 12. R_{IR} and R_P increased linearly with operating time. The increasing rate of R_P, which is associated with activation and concentration overvoltages, is 0.15 mΩ cm^2 h^{-1}, smaller than that of R_{IR} (0.21 mΩ cm^2 h^{-1}), indicating that the gradual loss of electrical contact was the main reason of the degradation. In this durability test, distinct morphology change and coking were not detected inside of the porous Ni-ScSZ anode support (see Fig. 14).

The composition of the anode off-gas (reformate gas) of IRSOFC operating on wet palm-BDF (S/C = 3.5) at 800°C was monitored during the durability test. In Table 9, the composition of the anode off-gas and open circuit voltage (OCV) just before the durability test are listed together with the equilibrium gas composition estimated by HSC 5.1 software (Outokumpu Research Oy, Finland) and the theoretical electromotive force calculated by Nernst equation. Internal steam reforming of palm-BDF led to H$_2$ rich reformate gas, indicating that syngas as well as electricity can be obtained directly from BDF using high temperature SOFC. Measured concentrations of H$_2$, CO and CO$_2$ were close to the calculated values. The initial OCV, 0.93 V, was close to the thermodynamic value of 0.94 V. The existence of methane and ethylene indicates that the internal steam reforming has not reached equilibrium. Especially, ethylene is well known as a precursor of carbon deposition (Yoon et al., 2008, 2009).

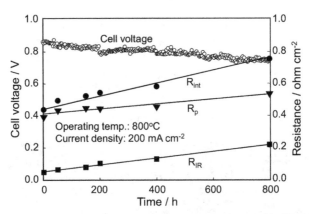

Fig. 12. The result of long term test of IRSOFC operating on wet palm-BDF (S/C = 3.5) at 800 °C, showing cell voltage, total ohmic resistance R_{IR}, total polarization resistance R_P and internal resistance of the cell R_{int} during the test.

	Outlet gas concentration (dry basis) / %					OCV / V
	H$_2$	CO	CO$_2$	CH$_4$	C$_2$H$_4$	
Measured	65.3	14.9	15.0	2.3	2.5	0.93
Calculated	70.8	14.4	14.8	0	0	0.94

Table 9. Anode off-gas composition and OCV for an IRSOFC operating on palm-BDF operated at 800°C just before the durability test shown in Fig. 12.

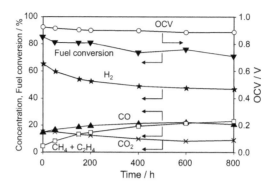

Fig. 13. Time dependence of OCV, fuel conversion and anode off-gas composition under open-circuit condition during the long term test of an IRSOFC operating on wet palm-BDF (S/C = 3.5) at 800°C shown in Fig. 12.

Figure 13 shows the concentrations of the gaseous species in the anode off-gas, fuel conversion and OCV during the long term test of Fig. 12. Although OCV was stable over the test, H_2 production rapidly degraded within the first 400 h. At least 800 h was necessary for the stabilization of the internal steam reforming of palm-BDF.

After stopping the supply of palm-BDF, cell temperature was decreased to room temperature under the thorough N_2 purging of the anode compartment. FESEM images of the (a) surface and (b) cross section of the anode support after the long term test are shown in Fig. 14. Carbon deposition occurred only on the surface of the anode which is most susceptible to coking due to the highest concentration of long chain hydrocarbons or lowest S/C, whereas inside of the porous anode support there was no coke. The occurrence of electrochemical consumption of H_2 leading to an increase in local S/C and direct electrochemical consumption of carbon may prevent coking inside of the anode. Carbon deposited on the anode surface can push up the current collector mesh attached to the anode surface resulting in the R_{IR} increase shown in Fig. 12.

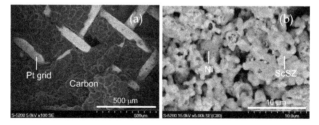

Fig. 14. FESEM images of (a) surface and (b) inside of the porous anode support after 800 h test of IRSOFC operating on wet palm-BDF (S/C = 3.5) at 0.2 A cm^{-2} and 800°C.

4.3 Problems to be solved for the realization of *Bio*-SOFCs

The feasibility of an IRSOFC running on low-grade biofuels has been demonstrated in the previous research (Shiratori et al., 2010a, 2010b, 2011) using anode-supported button cells. However, as illustrated in Fig. 15, in the real SOFC system a strong temperature gradient along gas flow direction exists and can cause cell fracture.

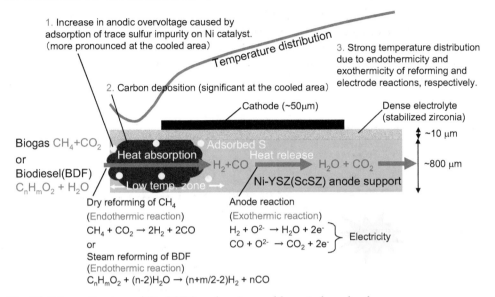

1. Increase in anodic overvoltage caused by adsorption of trace sulfur impurity on Ni catalyst. (more pronounced at the cooled area)

Temperature distribution

2. Carbon deposition (significant at the cooled area)

3. Strong temperature distribution due to endothermicity and exothermicity of reforming and electrode reactions, respectively.

Cathode (~50μm)

Dense electrolyte (stabilized zirconia)

~10 μm

Biogas $CH_4 + CO_2$
or
Biodiesel(BDF)
$C_nH_mO_2 + H_2O$

Heat absorption
Adsorbed S
Heat release
$H_2 + CO$ → $H_2O + CO_2$

Low temp. zone →
Ni-YSZ(ScSZ) anode support

~800 μm

Dry reforming of CH_4
(Endothermic reaction)
$CH_4 + CO_2 \rightarrow 2H_2 + 2CO$
or
Steam reforming of BDF
(Endothermic reaction)
$C_nH_mO_2 + (n-2)H_2O \rightarrow (n+m/2-2)H_2 + nCO$

Anode reaction
(Exothermic reaction)
$H_2 + O^{2-} \rightarrow H_2O + 2e^-$
$CO + O^{2-} \rightarrow CO_2 + 2e^-$ } Electricity

Fig. 15. Schematic view of *Bio*-SOFC and major problems to be solved.

The area near the fuel inlet is cooled down due to the strong endothermicity of reforming reactions (dry and steam reforming reactions of hydrocarbons), whereas cell temperature is gradually elevated toward the gas outlet by the exothermic electrochemical reactions. It is thermodynamically expected that impurity poisoning and carbon deposition are more significant at the cooled area.

4.3.1 Impurity poisoning (a case study of biogas operation)

Generally biofuels including biogas and BDFs contain several kinds of impurities such as sulfur compounds. According to thermochemical calculations (Haga et al., 2008), sulfur compounds exist as H_2S at SOFC operational temperatures in equilibrium. Here, in order to investigate the influence of a fuel impurity, 1 ppm H_2S was mixed with simulated biogas mixture (CH_4/CO_2 = 1.5) during the galvanostatic measurement under 200 mA cm^{-2}. CO yield and selectivity were measured simultaneously with the electrochemical measurements. CO yield and selectivity are defined as

$$CO \text{ yield} = v_{CO} / (f_{CH4} + f_{CO2}) \qquad (6)$$

$$CO \text{ selectivity} = v_{CO} / (v_{CH4} + v_{CO2}) \qquad (7)$$

where v_{CO} is CO formation rate, and f_{CH4} and f_{CO2} are feeding rates of CH_4 and CO_2, respectively, and v_{CH4} and v_{CO2} are consumption rates of CH_4 and CO_2, respectively. v_{CO}, v_{CH4} and v_{CO2} were estimated from the results of the exhaust gas analysis. The results of H_2S poisoning test at 1000 °C are summarized in Fig. 16.

In this experiment, an electrolyte-supported cell was used to measure anodic overvoltage separately from cathodic overvoltage using a Pt reference electrode (Shiratori, 2008). The horizontal axis indicates the time after starting poisoning. As shown in Fig. 16a even if 1 ppm H_2S was included in biogas, a cell voltage of about 1 V was stable during 20 h operation while voltage drop of about 100 mV (9 % of initial cell voltage) occurred in the first 1 h. The voltage drop was caused by increase in anodic overvoltage. Just after starting 1 ppm H_2S poisoning, anodic overvoltage grew to be 3.6 times larger compared to the initial value, which would be due to sulfur surface coverage of Ni catalysts. On the other hand, the anode-side IR drop did not change by the H_2S poisoning.

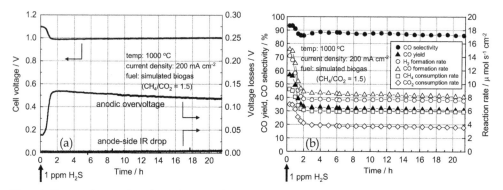

Fig. 16. Electrochemical and catalytic performance of Ni-ScSZ anode during 1 ppm H_2S poisoning measured at 1000 °C under 200 mA cm^{-2}; (a) cell voltage, anodic overvoltage (IR free) and anode-side IR drop and (b) reaction rates, CO yield and CO selectivity for internal dry reforming of methane. Simulated biogas mixture, CH_4/CO_2 = 1.5, was fed as a fuel (Shiratori et al., 2008).

The reaction rates of internal reforming, CO yield and CO selectivity during 1 ppm H_2S poisoning test are plotted on Fig. 16b. Rapid deactivation of the reforming reaction was observed within 2 h after starting H_2S poisoning resulting in about a 40 % decrease in reaction rates and CO yield. On the other hand, CO selectivity was less sensitive to H_2S contamination (only 7 % decrease). After the initial deactivation of the catalytic activity, a quasi-stable state appeared. Fig. 17 schematically depicts the degradation mechanism of an IRSOFC operating on biogas under H_2S contamination. Sulfur species chemisorbed on Ni catalyst not only deactivate reforming reaction but also block the triple phase boundary (TPB) which is the active reaction region for electrochemical oxidation of produced H_2 and CO. Deactivation of reforming reaction causes deficiency of H_2 and CO, and blockage of TPB sites causes an increase in local current density. These phenomena would appear as an initial increase in anodic overvoltage (or initial voltage drop). Cell voltage, CO yield and CO selectivity completely recovered within 4h after stopping H_2S poisoning, indicating that H_2S poisoning caused by adsorption of sulfur species is a reversible process and 1 ppm level H_2S contamination is not fatal for the operation of Bio-SOFC at 1000 °C.

On the contrary, at 800 °C, 1 ppm H_2S was more detrimental. Cell voltage and internal reforming rates kept on decreasing during 22 h of H_2S poisoning. Voltage drop of about 170 mV (20 % of initial cell voltage) and 80 % decrease in the reaction rates occurred without the quasi-stable state as in the case of 1000 °C testing. The results summarized in Table 10 indicate that higher-grade desulfurization is required for lower operating temperatures, suggesting that the effect of impurity poisoning will become more detrimental at the cooled area in the cell (see Fig. 15).

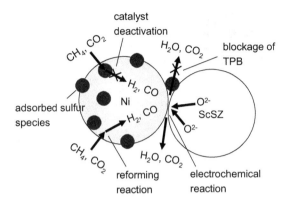

Fig. 17. Possible degradation mechanism of IRSOFC operating on biogas caused by H_2S contamination of biogas (Shiratori et al., 2008).

Temp.	Cell voltage	IR loss	Syngas production	Degradation behavior
1000 °C (Electrolyte-supported)	9 % decrease (100 mV)	not changed	40 % decrease	initial voltage drop followed by quasi-stable state
800 °C (Anode-supported)	20 % decrease (170 mV)	not changed	80 % decrease	continuous degradation

Table 10. Impact of 1 ppm H_2S contamination of biogas on the performance of IRSOFC running on biogas deduced by the 22 h poisoning test. Fuel: simulated biogas mixture (CH_4/CO_2 = 1.5), Cell: Ni-ScSZ/ScSZ/LSM-ScSZ (Shiratori et al., 2009a).

4.3.2 Strong temperature gradient (a case study of biogas operation)

Ni-ScSZ/ScSZ/LSM-ScSZ square-shaped cells with an area of 25 cm² simulating a real SOFC were used for the evaluation of thermomechanical reliability of a *Bio*-SOFC. When the simulated biogas (CH_4/CO_2 = 1.5) was directly fed to the square-shaped cell, long term operation could not be performed at 800 °C (Shiratori et al., 2010b). This is due to a locally-decreased cell temperature caused by the strong endothermicity of internal reforming (Fig. 15).

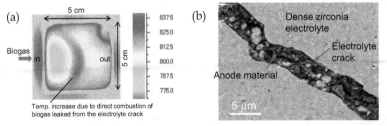

Fig. 18. (a) The temperature distribution caused by internal dry reforming of methane and (b) resulting cell fracture detected by FESEM, showing brittleness of electrolyte thin film sintered on the anode support versus direct supply of hydrocarbon fuel.

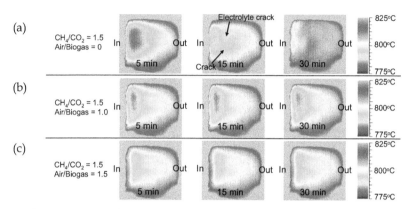

Fig. 19. The temperature distribution in the Ni-ScSZ anode support when simulated biogases with $CH_4/CO_2 = 1.5$ were fed directly with a furnace temperature of 800 °C under open circuit condition; (a) Air/Biogas =0, (b) 1.0 and (c) 1.5.

To estimate the endothermicity of reaction 1 in Table 7, an equimolar mixture of CH_4 and CO_2 was supplied to the square-shaped anode-supported half cell, and the temperature distribution generated in the half cell was measured from the electrolyte side by thermography. As shown in Fig. 18a, a large temperature gradient was generated at the fuel inlet side, and the temperature around the cooled area considerably increased within 20 min, indicating the formation of an electrolyte crack (Fig. 18b) from which biogas leaked and burned. When simulated biogas with $CH_4/CO_2 = 1.5$ was used as a fuel, at a furnace temperature of 800 °C, as shown in Fig. 19a, a strong temperature gradient formed in the half cell and led to an electrolyte crack within 15 min after starting the supply of simulated biogas. The temperature gradient became more moderate with air addition to the biogas, as can be seen in Fig. 19b and c, for air/biogas of 1.0 and 1.5, respectively, at the same furnace temperature.

	Reaction		Reaction enthalpy at 1100 K / kJ mol^{-1}
1	$CH_4 + CO_2 \rightarrow 2H_2 + 2CO$	Dry reforming of methane	259
2	$CH_4 + 1/2O_2 \rightarrow 2H_2 + CO$	Partial oxidation of methane	-23
3	$CH_4 + 2O_2 \rightarrow 2H_2O + CO_2$	Complete oxidation of methane	-802
4	$CH_4 + H_2O \rightarrow 3H_2 + CO$	Steam reforming of methane	226
5	$CH_4 \rightarrow C + 2H_2$	Methane cracking	90
6	$2CO \rightarrow C + CO_2$	Boudouard reaction	-170
7	$CO + H_2O \rightarrow H_2 + CO_2$	Water-gas-shift reaction	-34
8	$CO + 1/2O_2 \rightarrow CO_2$	Oxidation of carbon monoxide	-282
9	$H_2 + 1/2O_2 \rightarrow H_2O$	Oxidation of hydrogen	-248

Table 11. Main reactions involved in the reformig of air-mixed biogas and their reaction enthalpies at 1100 K (D.R. Lide (Ed.), 2009).

The above results indicate that cell performance can be stabilized by addition of air to biogas as a result of the exothermicity of partial oxidation of CH_4. Here, heat absorption accompanied by the internal reforming of air-mixed biogas was calculated thermodynamically for the feed of 1 kmol C considering the reactions listed in Table 11. Reactions 1-3 are the predominant reactions determining the amount of heat absorption. Reaction 1 is dry reforming of CH_4, producing H_2 and CO, which is a strong endothermic reaction. Reactions 2 and 3 are oxidation reactions of CH_4 which can contribute to the suppression of the strong endothermicity of reaction 1. The results of the thermodynamic calculation are plotted on Fig. 20.

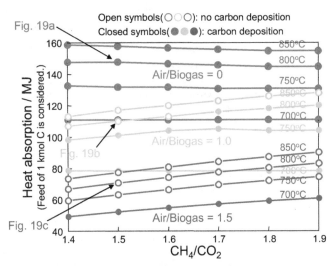

Fig. 20. Endothermicity of reforming of air-mixed biogases with Air/Biogas = 0, 1.0 and 1.5.

Fig. 20 suggests that heat absorption or endothermicity can be suppressed by increasing the air mixing ratio or by decreasing the operational temperature. A homogeneous temperature distribution obtained for Air/Biogas = 1.5 (Fig. 19c) as compared to Air/Biogas = 0 (Fig. 19a) is due to a 48 % lower heat absorption. Here, it is noted that the addition of excess air will result in the reduction of reforming efficiency, and lower operating temperature tends to cause carbon formation. The optimum Air/Biogas and operating temperature must be determined carefully, considering these points.

4.3.3 Carbon deposition

4.3.3.1 Biogas

The result of a long term test with a square-shaped cell is shown in Fig. 21a (Shiratori et al., 2010b). A degradation rate of 2.6 %/1000 h (200-500 h) was achieved by feeding air-mixed simulated biogas. The reasons for the degradation were a decrease in OCV (contribution ratio = 38 %) caused by incomplete gas sealing around the cell and increases in IR loss (34 %) and overvoltage (29 %) caused by insufficient electrical contact between the anode support and current collector. These voltage losses are related to technical difficulties of single cell testing with a square-shaped cell.

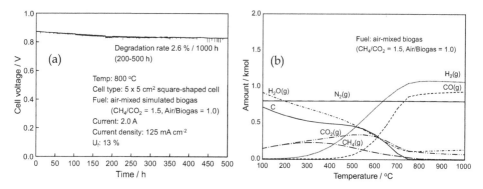

Fig. 21. Long-term test of an IRSOFC operating on biogas using a 5 x 5 cm² square-shaped cell: (a) galvanostatic measurement of cell voltage in air-mixed simulated biogas (CH_4/CO_2 = 1.5, Air/Biogas = 1) under U_f = 13 % at 800 °C and (b) the equilibrium gas composition calculated by HSC 5.1 software (Outokumpu Research Oy, Finland) for 0.6 kmol CH_4 + 0.4 kmol CO_2 + 0.2 kmol O_2 + 0.8 kmol N_2.

As shown in Fig. 21b, at 800 °C (operational temperature in the present study) direct feeding of air-mixed biogas (CH_4/CO_2 = 1.5, Air/Biogas = 1) thermodynamically does not cause carbon deposition. The safe temperature region for this fuel is above 750 °C, and a decrease in cell temperature over 50 K from the operational temperature of 800 °C may cause carbon deposition. After the 500 h durability test, carbon deposition was observed only at the fuel inlet side where the anode-support may be cooled down by more than 50 K as a result of the endothermic reforming reaction (Shiratori et al., 2010b). According to the thermodynamic calculation shown in Fig. 20, only 26% reduction of heat absorption is expected by the addition of an equimolar amount of air to biogas (Air/Biogas = 1.0). These results suggest that Air/Biogas = 1.0 is a thermodynamically safe composition if the temperature of 800 °C is kept anywhere in the anode-support, however taking the temperature gradient caused by the internal reforming into account, Air/Biogas = 1.0 leads to a decrease in the cell temperature at the fuel inlet side by more than 50 K, but is insufficient, Air/Biogas ratio higher than 1.0 is required practically.

4.3.3.2 Biodiesel fuels (BDFs)

Figure 22 shows the results of galvanostatic measurements at three different operating temperatures for palm-, jatropha- and soybean-BDFs under the condition of 200 mA cm⁻² and S/C = 3.5. Stable voltage with little oscillation was obtained only for palm-BDF at 800 °C. In contrast, jatropha- and soybean-BDFs resulted in unstable cell voltage. Especially, at 700 °C, cell voltage for the SOFCs operating on jatropha- and soybean-BDFs dropped abruptly within 40 h and 47 h, respectively. No severe coking was observed at 800 °C in the case of palm-BDF, whereas jatropha- and soybean-BDFs led to significant amount of carbon on the anode surface. Carbon deposition tended to be more significant at lower operating temperatures and at higher content of unsaturated FAMEs in BDF.

Figure 23 shows the anode surface after the operation with real biodiesels at 800 °C. Nearly no carbon was observed at this temperature in the case of palm-BDF, whereas jatropha- and soybean-BDFs led to significant amount of carbon on the anode surface. Carbon deposition tended to be more significant for the fuels with a higher degree of unsaturation (Nahar,

2010). The rapid degradation is due to the carbon deposition promoted with a higher degree of unsaturation which can cover a portion of active reaction sites or block open pores.

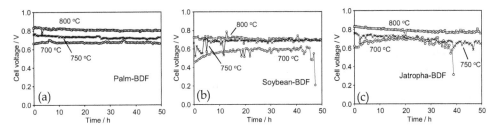

Fig. 22. Cell voltage of IRSOFCs operating on (a) palm-, (b) jatropha- and (c) soybean-BDFs at different temperatures under the condition of S/C = 3.5 and 200 mA cm⁻².

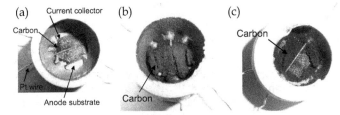

Fig. 23. Pictures of the anode after the 50 h tests of IRSOFCs operating on BDFs at 800 ºC shown in Fig. 22; (a), (b) and (c) are for palm-, jatropha- and soybean-BDFs, respectively.

5. Conclusion

The fuel flexibility of SOFCs has been demonstrated in this study with biofuels, biogas and biodiesel fuels (BDFs). In the course of this study, roadblocks for the realization of an internal reforming SOFC (IRSOFC) operating on biofuels, *Bio*-SOFC, which is promising candidate for a distributed generator in a coming carbon-neutral society, were uncovered. A biogas-fueled SOFC can be operated internal reforming mode if an adequate amount of air is added to the biogas. To suppress carbon deposition and thermal stress caused by the internal reforming reaction, Air/Biogas > 1.0 is required. Although the feasibility of an IRSOFC running on biogas has been demonstrated in the previous research, there is an urgent need to collect more practical data using a real stack. On the other hand, performance of an IRSOFC operating on BDFs has also been evaluated. The deactivation of the anode was accompanied by significant carbon deposition on the Ni-based anode material, which occurs more for lower operating temperature and higher concentration of unsaturated fatty acid methyl esters (FAMEs) in BDF. Only the IRSOFC operating on palm-BDF with the lowest degree of unsaturation operated at 800 ºC exhibited stable performance without severe coking. The concentration of unsaturated FAMEs in BDF is quite an important factor to determine SOFC performance, and therefore it should be controlled carefully. In terms of fuel injection performance, a higher concentration of unsaturated components resulting in lower viscosity of BDFs is preferable, however significant carbon deposition will occur as mentioned above.

Our final goal is to operate an IRSOFC using low grade biofuels. BDF derived from waste cooking oil which has a similar chemical composition to that of palm-BDF is quite an attractive candidate for SOFC operation, although the fuel composition fluctuates depending on the origin of the oil and kind of food cooked by the oil. On the other hand, south-east Asian nations near the Mekong basin are very interested in the efficient use of their abundant fishery resources as alternative fuels. BDF derived from discarded catfish oil is one of the most attractive alternative fuels, however it contains not only FAMEs with high number of carbon atom between 20 and 24 in a molecule, but also several trace impurities which promote carbon deposition. In future work, the dependence of IRSOFC performance on composition of BDFs will be investigated.

6. Acknowledgement

The authors are deeply grateful to Saga prefecture for their financial support to carry out this research and Tosu Kankyo Kaihatsu Inc. for their cooperation in operating the methane fermentation reactor.

7. References

Byham, S.; Spendelow, J.; Martin, K. E.; Ho, D. L.; Marcinkoski, J. & Papageorgopoulos, D. (2010). U.S. Department of energy activities in supporting fuel cell technologies for CHP and APU applications. in 9th European SOFC Forum, Connor, P.; Irvine, J.; Cassidy, M.; Savaniu, C.; Smith, M. & Knowles, S. Editors, pp.2.1-2.11, European Fuel Cell Forum Proceedings Series, Switzerland, 2010.

Föger, K. (2010). Commercialisation of CFCL's residantial power station BluGen. in 9th European SOFC Forum, Connor, P.; Irvine, J.; Cassidy, M.; Savaniu, C.; Smith, M. & Knowles, S. Editors, pp.2.22-2.28, European Fuel Cell Forum Proceedings Series, Switzerland, 2010.

Gardner, F.J.; Day, M.J.; Brandon, N.P.; Pashley, M.N. & Cassidy, M. (2000). SOFC technology development at Rolls-Royce. Journal of Power Sources, 86:122–129, ISSN 0378-7753.

Girona, K.; Laurencin, J.; Petitjean, M.; Fouletier, J. & Lefebvre-Joud, F. (2009). SOFC running on biogas: identification and experimental validation of "Safe" operating conditions. ECS Transactions, 25(2):1041-1050, ISSN 1938-5862.

Haberman, B. A.; Baca, C. M. & Ohrn, T. R. (2011). IP-SOFC performance measurement and prediction. ECS Transactions, 35(1):451-464, ISSN 1938-5862.

Haga, K.; Adachi, S.; Shiratori, Y.; Itoh, K. & Sasaki, K. (2008). Poisoning of SOFC anodes by various fuel impurities. Solid State Ionics, 179:1427-1431, ISSN 0167-2738.

Hosoi, K.; Ito, M. & Fukae, M. (2011). Status of national project for SOFC development in Japan. ECS Transactions, 35(1):11-18, ISSN 1938-5862.

Huang, P. & Ghezel-Ayagh, H. (2011). SOFC module material development at fuel cell energy. ECS Transactions, 35(1):2631-2638, ISSN 1938-5862.

Iida, T.; Kawano, M.; Matsui, T.; Kikuchi, R. & Eguchi, K. (2007). Internal reforming of SOFCs: carbon deposition on fuel electrode and subsequent deterioration of cell. Journal of Electrochemical Society, 154(2):B234–B241, ISSN 1945-7111.

Ke, K.; Gunji, A.; Mori, H.; Tsuchida, S.; Takahashi, H.; Ukai, K.; Mizutani, Y.; Sumi, H.; Yokoyama, M. & Waki, K. (2007). Effect of oxide on carbon deposition behavior of CH_4 fuel on Ni/ScSZ cermet anode in high temperature SOFCs. Solid State Ionics, 177:541–547, ISSN 0167-2738.

Kim, H.; Park, S.; Vohs, J.M. & Gorte R.J. (2001). Direct oxidation of liquid fuels in a solid oxide fuel cell. *Journal of Electrochemical Society*, 148(7):A693–A695, ISSN 1945-7111.

Kishimoto, H.; Yamaji, K.; Horita, T.; Xiong, Y.; Sakai, N.; Brito, M. & Yokokawa, H. (2007). Feasibility of liquid hydrocarbon fuels for SOFC with Ni-ScSZ anode. *Journal of Power Sources*, 172:67–71, ISSN 0378-7753.

Komatsu, Y.; Kimijima, S. & Szmyd, J.S. (2009). A performance analysis of a solid oxide fuel cell-micro gas turbine hybrid system using biogas. *ECS Transactions*, 25:1061-1070, ISSN 1938-5862.

Lanzini, A. & Leone, P. (2010). Experimental investigation of direct internal reforming of biogas in solid oxide fuel cells. *International Journal of Hydrogen Energy*, 35:2463-2476, ISSN 0360-3199.

Leah, R.; Bone, A.; Selcuk, A.; Corcoran, D.; Lankin, M.; Dehaney-Steven, Z.; Selby, M. & Whalen, P. (2011). Development of highly robust, volume-manufacturable metal-supported SOFCs for operation below 600°C. *ECS Transactions*, 35(1):351-367, ISSN 1938-5862.

Leone, P.; Lanzini, A.; Santarelli, M.; Cali, M.; Sagnelli, F.; Boulanger, A.; Scaletta, A. & Zitella, P. (2010). Methane-free biogas for direct feeding of solid oxide fuel cells. *Journal of Power Sources*, 195:239-248, ISSN 0378-7753.

Lide, D.R. (Ed.). (2008). *CRC Handbook of Chemistry and Physics*, Taylor & Francis Inc., ISBN 978-1-4200-6679-1, USA.

Mai, A.; Iwanschitz, B.; Weissen, U.; Denzler, R.; Haberstock, D.; Nerlich, V. & Schuler, A. (2011). Status of Hexis' SOFC stack development and the Galileo 1000 N micro-CHP system. *ECS Transactions*, 35(1):87-95, ISSN 1938-5862.

Nahar, N. (2010). Hydrogen Rich Gas Production by the autothermal reforming of biodiesel (FAME) for utilization in the solid-oxide fuel cells: A Thermodynamic Analysis. *International Journal of Hydrogen Energy*, 35:8891-8911, ISSN 0360-3199.

Park, S.; Cracium, R.; Vohs, J.M. & Gorte, R.J. (1999). Direct oxidation of hydrocarbons in a solid oxide fuel cell: I. methane oxidation. *Journal of Electrochemical Society*, 146(10):3603–3605, ISSN 1945-7111.

Payne, R.; Love, J. & Kah, M. (2011). CFCL's BlueGen product. *ECS Transactions*, 35(1):81-85, ISSN 1938-5862.

Sasaki, K. & Teraoka, Y. (2003). Equilibria in fuel cell gases: I. equilibrium compositions and reforming conditions. *Journal of Electrochemical Society*, 150(7):A878–A884, ISSN 1945-7111.

Sasaki, K.; Haga, K.; Yoshizumi, T.; Minematsu, D.; Yuki, E.; Liu, R.-R.; Uryu, C.; Oshima, T.; Taniguchi, S.; Shiratori, Y.; & Ito, K. (2011). Impurity poisoning of SOFCs. *ECS Transactions*, 35(1):2805-2814, ISSN 2151-2051.

Shiratori, Y.; Oshima, T. & Sasaki, K. (2008). Feasibility of direct-biogas SOFC. *International Journal of Hydrogen Energy*, 33:6316–6321, ISSN 0360-3199.

Shiratori, Y.; Ijichi, T.; Oshima, T. & Sasaki, K. (2009a). Generation of electricity from organic bio-wastes using Solid Oxide Fuel Cell. *ECS Transactions*, 25(2): 1051-1060, ISSN 1938-5862.

Shiratori, Y.; Kazunari, S.; Tran, Q.T. & Huynh, Q. (2009b). Application of biofuels to solid oxide fuel cell. *Proc. of the 2009 International Forum on Strategic Technologies*, pp. 89-93, Ho Chi Minh, Viet Nam, October 21-23, 2009.

Shiratori, Y.; Ijichi, T.; Oshima, T. & Sasaki, K. (2010a). Internal reforming SOFC running on biogas. *International Journal of Hydrogen Energy*, 35:7905–7912, ISSN 0360-3199.

Shiratori, Y.; Ijichi, T.; Oshima, T. & Sasaki, K. (2010b). Performance of internal reforming SOFC running on biogas. in *9th European SOFC Forum*, Connor, P.; Irvine, J.;

Cassidy, M.; Savaniu, C.; Smith, M. & Knowles, S. editors, pp.4-77-4-87, European Fuel Cell Forum Proceedings Series, Switzerland, 2010.

Shiratori, Y.; Tran, Q.T.; Takahashi, Y. & Sasaki, K. (2011). Application of biofuels to solid oxide fuel cell. *ECS Transactions*, 35(1):2641-2651, ISSN 2151-2051.

Staniforth, J. & Kendall, K. (1998). Biogas powering a small tubular solid oxide fuel cell. *Journal of Power Sources*, 71(1-2):275-277, ISSN 0378-7753.

Staniforth, J. & Kendall, K. (2000). Cannock landfill gas powering a small tubular solid oxide fuel cell - a case study. *Journal of Power Sources*, 86:401-403, ISSN 0378-7753.

Steele, BCH & Heinzel A. (2001). Materials for Fuel-cell Technologies. *Nature*, 414:345–352, ISSN 0028-0836.

Stöver, D.; Buchkremer, H.P. & Huijsmans, J.P.P. (2003) MEA/cell preparation methods: Europe/USA, In: *Handbook of Fuel Cells-Fundamentals, Technology and Applications*, Vielstich, W.; Lamm, A. & Gasteiger, H. A. (Eds.), 1015-1031, John Wiley & Sons Ltd., ISBN 0-471-49926-9, England.

Tran, Q.T.; Shiratori, Y. & Sasaki, K. (2011). Feasibility of palm-biodiesel fuel for internal reforming Solid Oxide Fuel Cells. *International Journal of Energy Research* (submitted), ISSN 1099-114X.

Van herle, J.; Membrez, Y. & Bucheli, O. (2004a). Biogas as a fuel source for SOFC co-generators. *Journal of Power Sources*, 127:300–312, ISSN 0378-7753.

Van herle, J.; Maréchal, F.; Leuenberger, S.; Membrez, Y.; Bucheli, O. & Favrat, D. (2004b). Process flow model of solid oxide fuel cell system supplied with sewage biogas. *Journal of Power Sources*, 131:127-141, ISSN 0378-7753.

Xuan, J.; Leung, M.K.H.; Leung, D.Y.C. & Ni, M. (2009) A review of biomass-derived fuel processors for fuel cell systems. *Renewable and Sustainable Energy Reviews*, 13:1301-1313, ISSN 1364-0321.

Yentekakis, I.V. (2006). Open- and closed-circuit study of an intermediate temperature SOFC directly fueled with simulated biogas mixtures. *Journal of Power Sources*, 160:422–425, ISSN 0378-7753.

Yoon, S.; Kang, I. & Bae, J. (2008). Effects of ethylene on carbon formation in diesel autothermal reforming. *International Journal of Hydrogen Energy*, 33:4780-4788, ISSN 0360-3199.

Yoon, S.; Kang, I. & Bae, J. (2009). Suppression of ethylene-induced carbon deposition in diesel autothermal reforming. *International Journal of Hydrogen Energy*, 34:1844–1851, ISSN 0360-3199.

Yoshida, S.; Kabata, T.; Nishiura, M.; Koga, S.; Tomida, K.; Miyamoto, K.; Teramoto, Y.; Matake, N.; Tsukuda, H.; Suemori, S.; Ando, Y. & Kobayashi, Y. (2011). Development of the SOFC-GT combined cycle system with tubular type cell stack. *ECS Transactions*, 35(1):105-111, ISSN 1938-5862.

Yuki, E.; Haga, K.; Shiratori, Y.; Ito, K. & Sasaki K. (2009). Co-poisoning effects by sulfur impurities and hydrocarbons in SOFCs. *Proc. 18th Symposium on Solid Oxide Fuel Cells in Japan*, pp.104-107, Tokyo, Japan, December 2009.

Zhou, Z.F.; Gallo, C.; Pargue, M.B.; Schobert, H. & Lvov S.N. (2004). Direct oxidation of Jet fuels and pennsylvania crude oil in a solid oxide fuel cell. *Journal of Power Sources*, 133:181–187, ISSN 0378-7753.

Zhou, Z.F.; Kumar, R.; Thakur, S.T.; Rudnick, L.R.; Schobert, H. & Lvov, S.N. (2007). Direct oxidation of waste vegetable oil in solid-oxide fuel cells. *Journal of Power Sources*, 171:856-860, ISSN 0378-7753.

Züttel, A. (2008). Material for the hydrogen world. in *Ceramic materials in energy system for sustainable development*, Gauckler, L.J. Editor. Forum 2008 of the World Academy of Ceramics, ISBN 978-88-86538-50-3, pp.211-260, Italy, 2008.

8

Bioethanol-Fuelled Solid Oxide Fuel Cell System for Electrical Power Generation

Vorachatra Sukwattanajaroon[1], Suttichai Assabumrungrat[1],
Sumittra Charojrochkul[2], Navadol Laosiripojana[3] and
Worapon Kiatkittipong[4]
[1]Chulalongkorn University, Faculty of Engineering, Department of Chemical Engineering
[2]National Metal and Materials Technology Center
[3]King Mongkut's University of Technology Thonburi
[4]Silpakorn University, Department of Chemical Engineering
Thailand

1. Introduction

Tremendous consumption of energy to serve daily lives and economic activities has led to the critical problem of energy shortage since the current main energy sources rely on fossil fuels which are non-renewable. Therefore, efficient renewable energy sources need to be investigated and improved to replace or substitute the use of fossil fuels to alleviate environmental impacts while being sustainable. Biomass-derived fuels are recognized as promising alternatives among other renewable sources e.g. wind, solar, geothermal, hydropower, etc. This fuel can be produced from various available agricultural materials, hence there is no problem of feedstock supply. Instead, its use is beneficial for those countries having strong background in agriculture. In addition, this agro-based fuel can provide a CO_2-closed cycle as the CO_2 released from the fuel combustion can be redeemed with the CO_2 required for biomass growth. Bioethanol plays an important role as a promising renewable energy among other biofuels due to its useful properties such as high hydrogen content, non-toxicity, safety, ease of storage and handling (Ni et al., 2007). An efficient energy conversion system is required to maximize bioethanol fuel utilization to obtain a full performance. Combustion heat engines which are widely used nowadays have a low conversion efficiency of power production due to losses during multiple energy conversion stages as well as a low value of chemical energy of bioethanol represented by LHV or HHV compared to those of fossil fuels (C_6 hydrocarbons or above). Moreover, electrical energy efficiency produced from a combustion heat engine becomes even lower because of further losses from more energy conversion stages. Fuel cell technology is considered to be an interesting alternative for efficient energy conversion since it can directly convert chemical energy stored in the fuel into electrical energy via electrochemical reaction. Less energy is lost in the fuel cell operation and higher electrical efficiency can be obtained. However, the problems in using fuel cell technology such as short-life operating time, high manufacturing cost and impromptu infrastructure support are still issues to be tackled. The Solid Oxide Fuel Cell (SOFC), a type of fuel cells, is selected to be an electrical power generation unit fuelled by bioethanol because of its outstanding characteristics: ability to use low-cost catalyst, high temperature exhaust heat for cogeneration application, tolerance to some impurities e.g. CO and sulfur, internal reforming within the cell for reducing equipment cost, etc. For the SOFC system, bioethanol feed is heated up and reformed to hydrogen rich gas by the reformer before being introduced into the fuel cell at the anode side coupled with air feed at the cathode side for producing electricity. To achieve better performance from this process, it is necessary to consider every unit within the SOFC system. These units are investigated through their physical structure design and modification on the basis of worthwhile energy utilization in each unit and suitable energy allocation within the process to target an optimum energy management of the SOFC system. The objective of this chapter is to propose ideas and feasible approaches on how to improve the performance of bioethanol-fuelled SOFC systems by focusing on each essential unit modification in the process. Relevant useful approaches from other scientific literature reviews are included. The pros and cons in each proposed method are also discussed. Bioethanol pretreatment unit regarded as a significant unit compared to the other units for the process development is of particular focus in this chapter. The progressive work of our research on the efficiency improvement of the SOFC system with analytically appropriate selection of bioethanol pretreatment unit is presented. The simulation studies were conducted via experimental-verified SOFC model to predict the results under a frame of model assumptions. Performance assessment of the system in any scenario cases held the criteria of no external energy demand condition or $Q_{net} = 0$ to compare and identify the optimal operating conditions among those of bioethanol pretreatment units. The simulation results could initially guide the right pathway for practical industrial applications.

2. Bioethanol

Among various biomass-based fuel types such as bioethanol, biodiesel, bioglycerol, and biogas, bioethanol is considered a promising renewable energy compared to other biofuels. As shown in Table 1, the maximum amount of work from the fuel cell integrated with fuel processor system in comparison with five renewable fuels including n-octane represented as a gasoline characteristic are presented (Delsman et al., 2006). It was indicated that ethanol can offer the highest energy output (based on MJ/mol fuel) among the other renewable fuels (methanol, methane, ammonia, and hydrogen) except for n-octane. Furthermore, there are other outstanding advantages of bioethanol given by the following reasons. The production technologies of bioethanol are more mature and cheaper than those of biomethanol (Xuan et al., 2009). Biodiesel which is a popular alternative energy used in vehicle engines can be derived from ethanol (or purified bioethanol) reacted with vegetable oil via transesterification reaction. Biogas is a widely-used renewable power source because of many available feedstocks. It can be produced from several organic wastes by anaerobic biological fermentation. Consequently, it seems to be a promising renewable fuel but biogas mainly consists of methane and CO_2. Both gases have serious negative environmental impacts especially from methane. Methane can remain in atmosphere for 9-15 years and retains heat radiation of 20 times higher than CO_2 (U.S. Environmental Protection Agency). Furthermore, if the biogas is produced from non-agricultural wastes, e.g. cow and pig manure, it would bring this biogas production diverted from carbon-closed cycle. Hence, biogas should be produced and utilized in an effective way. Bioethanol production is mostly derived from biological fermentation using agro-based raw materials such as sucrose-containing crops, starchy materials, lignocellulosic biomass and agro-waste (Carlos & Oscar, 2007). In addition, the latest research reports that animal manure waste, waste paper, citrus peel waste, and municipal solid waste can be used as feedstock of bioethanol production by using saccharification and fermentation processes (Lal, 2008; Foyle et al., 2007; Wilkins et al., 2007).

Fuel	Maximum amount of work		
	MJ/mol Fuel	MJ/mol C in Fuel	MJ/mol H_2 via reforming
Methanol	-0.69	-0.69	-0.23
Ethanol	-1.31	-0.65	-0.22
n-Octane	-5.23	-0.65	-0.21
Ammonia	-0.33		-0.22
Methane	-0.8	-0.80	-0.20
Hydrogen	-0.23		-0.23

Table 1. Maximum amount of work for the conversion of fuels to electricity calculated at 298 K and 1 bar (Source: Delsman et al., 2006)

However, bioethanol fermentation is a complicated process. The overall process is schematically shown in Figure 1. It requires many steps of biomass feed conditioning or pretreatment which can be mainly divided into four techniques as follows (Magnusson, 2006):
- Mechanical techniques: biomass is milled or ground to reduce sizes of material,
- Chemical techniques: biomasses e.g. hemicelluloses and lignin are swelled or dissolved by acids, bases, and solvents to transform into pre-hydrolysis form,
- Mechanical-chemical techniques: a combined mechanical and chemical technique e.g. heat pretreatment with high-pressure of steam, and
- Biological techniques: biomass is digested by enzymes or micro-organisms.

Thereafter, the pre-conditioned biomass is biologically transformed into ethanol. This procedure is a key step to be accounted for increasing bioethanol productivity. The basic concept of reactor design is applied with enzymatic fermentation technology. Starting from a simple batch reactor, this is close to organic culture system environment but a batch culture envisages the limitation of enlarging bioethanol production scale. Afterward, semi-batch reactors combining the benefits of batch and continuous reactors are employed. It can offer a long lifetime of cell culture, higher ethanol and cell concentration (Frison & Memmert, 2002). Finally, a continuous flow reactor is applied with cell recycle operation to serve more bioethanol productivity requirement. Influent stream containing substrate, nutrients and culture medium is fed to an agitated bioreactor. The product is removed from the fermenter but the residues (cells and nutrients) are collected and recycled to the vessel. In addition, the concept of process integration is introduced to the bioethanol production application such as Separate Hydrolysis and Fermentation (SHF), Simultaneous Saccharification and Fermentation (SSF) and Direct Microbial Conversion (DMC) (Balat, 2011). In the last step, the obtained dilute ethanol is then purified to gain a desired ethanol concentration. These difficult procedures need to be further developed to reduce the complexity and enable the process to compete with the cheaper oil-derived fuel production.

Many researchers attempt to develop such a biotechnical bioethanol production to be cost-effective. Effective tools for the process evaluation such as thermo-economic, environmental indexes, process optimization and etc. are used to analyze the bioethanol production process as performance indicators to assist in the task of process design. Process integration is regarded as a significant approach since several production procedures are combined into a single unit. It can reduce production costs and provide a more intensive process. For example, the fermentation process integrated with membrane distillation (Gryta, 2001) involved the combination of tubular bioreactor and membrane distillation to synergistically enhance the yield of bioethanol without several units being required as for other common processes. A role of membrane distillation is to remove byproduct from the fermentation broth in bioreactor that can simultaneously forward glucose conversion to gain more ethanol. The objective of process integration is to have the energy requirement in procedures of bioethanol production to be less than the energy obtained from the bioethanol exploitation to utilize bioethanol effectively.

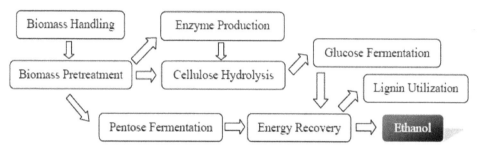

Fig. 1. Schematic diagram of bioethanol production process (Source: U.S. Department of Energy)

Bioethanol can be purified to anhydrous ethanol which is a useful chemical for various applications such as organic solvent, chemical reagent, reactant for biodiesel production, pharmaceutical formations, plastics, polishes and cosmetics industries (Kumar et al., 2010). However, in this chapter, the use of bioethanol is particularly focused on a role of renewable fuel. Application of bioethanol in term of fuel can be mainly divided by two directions:
- Direct combustion to gain thermal energy
- Reforming into hydrogen for clean energy production in a fuel cell

For conventional direct combustion, it seems to be less complicated but the fuel is utilized in an ineffective way because thermal energy accumulated in bioethanol is obviously lower than fossil fuel as shown in Table 2.

Fuel	Density (kg/l)	Caloric value at 20ºC (MJ/kg)	Caloric value (MJ/l)	Octane-number (RON)	Fuel-equivalence (l)
Petrol	0.76	42.7	32.45	92	1
Bioethanol	0.79	26.8	21.17	>100	0.65

Table 2. Properties of bioethanol in comparison with petrol (Source: Paul & Kemnitz, 2006)

Moreover, since water is the main constituent in bioethanol, the direct combustion of bioethanol is not possible. However, there is another effective way which is the conversion of bioethanol fuel into hydrogen rich gas. As presented in Table 3, the heating value of hydrogen is higher than that of ethanol (4.47 times). Therefore, the bioethanol reforming process for producing hydrogen is a promising pathway in term of upgrading fuel quality which can offer a higher performance for the SOFC system even in the combustion heat engine while the bioethanol fuel utilization can be conducted in an efficient way.

Fuel	Lower Heating Value (25 °C and 1atm)
Hydrogen	119.93 kJ/g
Methane	50.02 kJ/g
Gasoline	44.5 kJ/g
Diesel	42.5 kJ/g
Ethanol	26.82 kJ/g
Methanol	18.05 kJ/g

Table 3. Heating values of commonly-used fuels in comparison

Typically, there are three main reforming reactions for hydrogen production as described below:
- Steam reforming
- Partial oxidation
- Autothermal reforming

Selection of an appropriate operation mode depends on the individual objective. Ethanol steam reforming (ESR) (Reaction (1)) is a suitable choice for the SOFC system because this reaction can produce hydrogen at high yield. Although ESR consumes a great amount of heat due to its high endothermicity, heat released from the fuel cell is enough to supply the heat demand for the reaction. For the ethanol partial oxidation (EPOX) (Reaction (2)), it is appropriate for the process required less complexity and integration design. Since EPOX requires the fuel to be partly combusted with air and releases thermal energy as an exothermic reaction, heat and steam supply are not required (Vourliotakis et al., 2009). Nonetheless, this reaction is less selective to hydrogen compared to the former reaction.

Autothermal reforming (ATR) is a combination of the previous two reactions in order to improve the hydrogen selectivity with minimum heat supply. The steam to carbon molar ratio and air to carbon molar ratio are significant parameters to adjust the system to operate close to thermal neutral condition from the exothermic partial oxidation and endothermic steam reforming. Generally, this reaction formula is defined as Reaction (3) with the standard exothermic heat $\Delta H_{298K} = 50$ kJ/mol (Deluga et al., 2004). There is a scientific literature (Liguras et al., 2003) reporting the stoichiometric ratio of H_2O and O_2 of 1.78 and 0.61, respectively per mol of ethanol can carry out thermal neutrality as shown in Reaction (4) but the yield of hydrogen becomes a little lower.

$$C_2H_5OH + 3H_2O => 2CO_2 + 6H_2 \tag{1}$$

$$C_2H_5OH + 0.5O_2 => 2CO + 3H_2 \tag{2}$$

$$C_2H_5OH + 2H_2O + 0.5O_2 => 2CO_2 + 5H_2 \tag{3}$$

$$C_2H_5OH + 1.78H_2O + 0.61O_2 => 2CO_2 + 4.78H_2 \qquad (4)$$

3. Solid oxide fuel cell system fuelled by bioethanol

As mentioned earlier, utilization of bioethanol by being converted into H_2 for electrical power generation via SOFC is recognised. Thus, this section describes the fundamental process of an SOFC system fuelled by bioethanol and the criteria used to define the performance evaluation indicators of this SOFC system as follows:

3.1 Process description

The bioethanol-fuelled SOFC system basically consists of a bioethanol pretreatment unit, preheaters, reformer, fuel cell, and afterburner as illustrated in Figure 2. Bioethanol is purified in the pretreatment unit to achieve a specified ethanol concentration (25mol% ethanol, a suitable stoichiometric ratio for the ethanol steam reforming reaction in Reaction (1)). Then, the steam with a desired ethanol is fed to an external reformer operated under thermodynamic equilibrium condition. Ethanol steam reforming is selected for converting the raw materials into hydrogen rich gas. The reaction is assumed to occur isothermally in the reformer. Finally, the reformed influent stream is fed to the SOFC's anode chamber together with excess air (5 times) preheated and fed to the cathode chamber to produce electricity and thermal energy. The effluent steam containing residual fuel released from the fuel cell is combusted in the afterburner and heat from the fuel combustion is recovered to supply all the heat-demanding units i.e. preheaters, purification unit, and reformer. The final temperature of exhaust gas emitted to atmosphere is specified at 403K (Jamsak et al., 2007). The performance of the SOFC system can be simulated using Aspen Plus software.

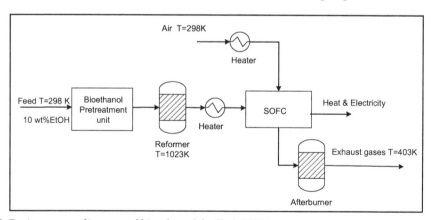

Fig. 2. Basic process diagram of bioethanol-fuelled SOFC system

3.2 Mathematical model

The SOFC model was adapted from the prior literature of Piroonlerkgul et al., 2008 to study the performance of SOFC system. From this model, a constant operating voltage along the cell length and isothermal condition were assumed. Only hydrogen oxidation was considered to react electrochemically within the fuel cell. Oxygen ion electrolyte type was chosen for the SOFC and its electrochemical reactions are described below:

$$0.5O_2 + 2e^- => O^{2-} \tag{5}$$

$$H_2 + O^{2-} => H_2O + 2e^- \tag{6}$$

The validation of this model was in a good agreement with experimental results (Zhao et al., 2005; Tao et al., 2005) at high hydrogen contents (hydrogen mole fraction = 0.97) and (Petruzzi et al., 2003) at low hydrogen contents (hydrogen mole fraction = 0.26). The materials used in the SOFC stack were YSZ, Ni-YSZ and LSM-YSZ for electrolyte, anode and cathode, respectively.

3.2.1 Electrochemical model

3.2.1.1 Open circuit voltage

The open circuit voltage (E) is formulated by the Nernst equation given in Eq. (7)

$$E = E_0 + \frac{RT}{F}\ln\left(\frac{P_{H_2}P_{O_2}^{1/2}}{P_{H_2O}}\right) \tag{7}$$

where F is Faraday constant (C mol^{-1}) and P_i is a partial pressure of component i.
The actual operating voltage (V) is less than the open circuit voltage (E) due to the presence of various polarizations. Three types of polarization are considered in this model: Ohmic, Activation, and Concentration polarizations as below:

$$V = E - \eta_{act} - \eta_{ohmic} - \eta_{conc} \tag{8}$$

3.2.1.2 Polarizations

3.2.1.2.1 Ohmic polarization

This ohmic polarization involves the resistance of both ions flowing in the electrolyte and electrons flowing through the electrodes. This resistance loss is regarded as a major loss in the SOFC stack given as:

$$\eta_{ohmic} = 2.99x10^{-11}\, iL\exp\left(\frac{10300}{T}\right) \tag{9}$$

where i is current density and L is thickness of anode electrode

3.2.1.2.2 Activation polarization

Activation polarization is caused by the loss of electrochemical reaction rate at the electrodes. An operation of SOFC at high temperature can reduce this polarization as the rate-determining step becomes faster. Normally, activation polarization region occurs in the low current density range. This polarization is defined by the Butler-Volmer equation.

$$i = i_0\left[\exp\left(\frac{\alpha zF\eta_{act}}{RT}\right) - \exp\left(-\frac{(1-\alpha)zF\eta_{act}}{RT}\right)\right] \tag{10}$$

The value of a and z were specified as 0.5 and 2 (Chan et al., 2001), respectively. Accordingly, the activation polarization at the anode and cathode sides can be arranged into another form as:

$$\eta_{act,j} = \frac{RT}{F}\sinh^{-1}\left(\frac{i}{2i_{0,j}}\right) \tag{11}$$

where j = anode, cathode

The exchange current density ($i_{o,j}$) for both the anode and cathode sides are expressed as follows:

$$i_{o,a} = \gamma_a\left(\frac{P_{H_2}}{P_{ref}}\right)\left(\frac{P_{H_2O}}{P_{ref}}\right)\exp\left(-\frac{E_{act,a}}{RT}\right) \tag{12}$$

$$i_{o,c} = \gamma_c\left(\frac{P_{O_2}}{P_{ref}}\right)^{0.25}\exp\left(-\frac{E_{act,c}}{RT}\right) \tag{13}$$

where γ_a and γ_c are pre-exponential factors for anode and cathode current densities, respectively.

3.2.1.2.3 Concentration polarization

This polarization arises from the difference in gas partial pressures in the porous electrode region due to slow mass transport. It can be estimated by Eqs. (14) and (15) for the anode and cathode sides, respectively.

$$\eta_{conc,a} = \frac{RT}{2F}\ln\left[\frac{\left(1+(RT/2F)(l_a/D_{a\,(eff)}P^I_{H_2O})i\right)}{(1-(RT/2F)(l_a/D_{a\,(eff)}P^I_{H_2})i}\right] \tag{14}$$

$$\eta_{conc,c} = \frac{RT}{4F}\ln\left[\frac{P^I_{O_2}}{(p_c/\delta_{O_2})-((p_c/\delta_{O_2})-p^I_{O_2})\exp\left[(RT/4F)(\delta_{O_2}l_c/D_{c(eff)}P_c)i\right]}\right] \tag{15}$$

where l_a and l_c are thicknesses of anode and cathode electrodes, respectively, while δ_{O2}, $D_{a\,(eff)}$ and $D_{c\,(eff)}$ are given by:

$$\delta_{O_2} = \frac{D_{O_2,k(eff)}}{D_{O_2,k(eff)}+D_{O_2-N_2(eff)}} \tag{16}$$

$$D_{a(eff)} = \left(\frac{P_{H_2O}}{P_a}\right)D_{H_2(eff)}+\left(\frac{P_{H_2}}{P_a}\right)D_{H_2O(eff)} \tag{17}$$

$$D_{c(eff)} = \frac{\xi}{n}\left(\frac{1}{D_{O_2,k}}+\frac{1}{D_{O_2-N_2}}\right) \tag{18}$$

$$\frac{1}{D_{H_2(eff)}} = \frac{\xi}{n}\left(\frac{1}{D_{H_2,k}}+\frac{1}{D_{H_2-H_2O}}\right) \tag{19}$$

$$\frac{1}{D_{H_2O(eff)}} = \frac{\xi}{n}\left(\frac{1}{D_{H_2O,k}} + \frac{1}{D_{H_2-H_2O}}\right)$$ (20)

The relationship between the effective diffusion parameter ($D_{(eff)}$) and ordinary diffusion parameter (D) can be described by:

$$D_{(eff)} = \frac{n}{\xi}D$$ (21)

where n is electrode porosity and ξ is electrode tortuosity. Assuming straight and round pores, the Knudsen diffusion parameter can be calculated by:

$$D_{A,k} = 9700\sqrt{\frac{T}{M_A}}$$ (22)

The binary ordinary diffusion parameter in a gas phase can be calculated using the Chapman-Enskog theory of prediction as below:

$$D_{A-B} = 1.8583\times10^{-3}\left(\frac{T^{3/2}((1/M_A)+(1/M_B))^{1/2}}{Po_{AB}^2\Omega_D}\right)$$ (23)

where σ_{AB} the characteristic length, M_i is molecular weight of gas i, and Ω_D is the collision integral. These parameters are given by:

$$\sigma_{AB} = \frac{\sigma_A + \sigma_B}{2}$$ (24)

$$\Omega_D = \frac{A}{T_k^B} + \frac{C}{\exp(DT_k)} + \frac{E}{\exp(FT_k)} + \frac{G}{\exp(HT_k)}$$ (25)

where the constants A to H are $A = 1.06036$, $B = 0.15610$, $C = 0.19300$, $D = 0.47635$, $E = 1.03587$, $F = 1.52996$, $G = 1.76474$, $H = 3.89411$.

3.3 Evaluation of process performance

The proposed bioethanol-fuelled SOFC system for electrical power generation needs to be evaluated together with any process design adjustments to obtain optimum performance. A number of criteria can be used to define the performance of the system, e.g. economic, 1st and 2nd laws of thermodynamics, environment, etc. Fundamentally, the overall performance evaluation of an SOFC system is defined in terms of electrical efficiency as below:

$$\eta_{elec,ov} = \frac{\text{Net electrical power output}}{\text{mol}_{Fuel}\cdot LHV_{Fuel}}$$ (26)

The definition of the above equation is energy efficiency based on 1st law of thermodynamics which initially accounts on an ideal energy conservation law. Fuel input term is referred to the lower heating value (LHV). In considering an energy loss from the system which is closer to actual condition, the definition of overall system efficiency is formulated as follows:

$$\eta_{elec,ov} = \frac{\text{Net electrical power output}}{\text{mol}_{Fuel} \cdot e^{o}_{Fuel}} \tag{27}$$

This equation is exergy efficiency which further takes the 2nd law of thermodynamic into account stated that entropy loss occurred in the system with highly irreversible process especially combustion process. The fuel input denominator in Eq. (27) is referred to the standard exergetic potential of fuel. In addition, the analysis in term of exergy can determine the location, source, and amount of actual thermodynamic inefficiencies in each unit. Profound understanding can be perceived from this analysis for solving the process problem correctly.

The criterion mainly considered in this chapter is no external energy demand condition. In the SOFC system, there are units having the roles of both energy consumption and generation. Before investigating and evaluating the system efficiency, energy consumed or generated from the units is allocated within the system until the overall system is under self-sufficient energy condition or $Q_{net} = 0$ as follows:

$$Q_{net} = Q_{generation} - Q_{demand} = 0 \tag{28}$$

where $Q_{generation}$ represents the heat from units which can generate thermal energy (SOFC and afterburner) while Q_{demand} expressed as the heat from units which consume heat (bioethanol pretreatment unit, heaters, and reformer). The system operated at such a condition can help allocate energy within the process effectively. The exhaust gas released to atmosphere is specified at 403 K (Jamsak et al., 2007). The consideration of $Q_{net} = 0$ associated with the process evaluation has led to the modified efficiency definition:

$$\eta_{elec,ov} = \frac{\text{Net electrical power output}}{\text{mol}_{Fuel} \cdot LHV_{Fuel} + \text{external heat demand}} \tag{29}$$

In case of incorporating a heat recovery unit such as combined heat and power (CHP) with the SOFC system, the definition of efficiency is adjusted to:

$$\eta_{elec,ov} = \frac{\text{Net electrical power output} + \text{exchanged thermal energy}}{\text{mol}_{Fuel} \cdot LHV_{Fuel} + \text{external heat demand}} \tag{30}$$

4. Process modification for improving performance of the SOFC system

The fundamental process of the bioethanol-fuelled SOFC system needs to be further developed to utilize bioethanol effectively and achieve higher electrical efficiency. In this chapter, the performance improvement of SOFC systems under consideration is based on selection for appropriate units. The possible units are structurally modified and evaluated for their energy consumption. The process modification of the SOFC system can be divided by two main scopes including adjusting the fuel cell module and improving the balance of plant.

4.1 Solid Oxide Fuel Cell
Originally, the Solid Oxide Fuel Cell (SOFC) is classified as a high-temperature fuel cell. Due to the demand for high cost materials and fabrication, the intermediate temperature solid

oxide fuel cell was later developed with the research into novel material technology and thin layer techniques applied in electrolyte and electrodes. Regarding the fuel cell geometry design, it is useful to differentiate the scope into macro and micro geometry configurations. The micro geometry covering the structures of anode, electrolyte, and cathode has direct effects on the electrochemical performance of the fuel cell. The heat transfer mechanisms of convection and conduction through heat exchange areas and the mass transport through active surface areas are influenced by the macro geometry (Nagel et al., 2008). Generally, primary structures of SOFC are tubular, planar, and monolithic as shown in Figures (3), (4), and (5), respectively. The SOFC structure of planar design is more compact than the tubular design and also offers higher ratio of power per volume (Pramuanjaroenkij et al., 2008). For the monolithic design, this SOFC design uses the similar concept with shell-and-tube heat exchanger. It combines the tri-layer of anode-electrolyte-cathode into a compact corrugated structure. This design can assist a thermal energy allocation exchanged between the flow channels and size of the fuel cell to become more compact with the corrugated self-supporting structure.

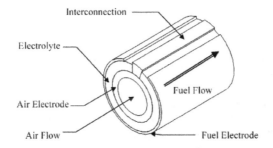

Fig. 3. Schematic of tubular SOFC (Source: Kakac et al., 2007)

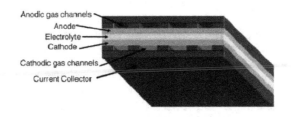

Fig. 4. Schematic of planar SOFC (Source: Bove & Ubertini, 2006)

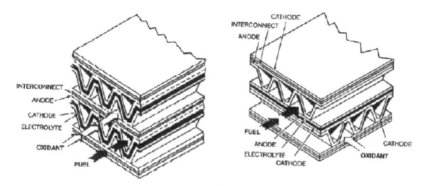

Fig. 5. Schematic of monolithic SOFC; (Left) coflow and (Right) cross flow (Source: Minh, 1993)

4.2 Balance of plant

There are essential units around the fuel cell as supporting units for the overall electrical power generation process. These units can be modified to utilize energy within their system units suitably. Sections in the balance of plant which are potential in improving the efficiency of SOFC system are described as follows:

4.2.1 Bioethanol pretreatment section

This section has a key role in improving the efficiency of the SOFC system. Originally, bioethanol has a low concentration in a range of 1-7 mol% (Shell et al., 2004; Cardona Alzate & Sanchez Toro, 2006; Roger et al., 1980; Buchholz et al., 1987). In our studies, 10wt% (4.16 mol%) ethanol was specified to represent the range of actual bioethanol concentration. These bioethanol compositions are unsuitable for feeding into the reformer operating under ethanol steam reforming reaction to produce hydrogen because of high water content. Unnecessary thermal energy is required to heat up surplus water within the reformer and the sizes of reformer are larger than necessary. Hence, the bioethanol pretreatment unit plays an important part to purify bioethanol feed into a desired concentration of 25mol%ethanol (46wt% ethanol). A selection of appropriate purification unit for bioethanol conditioning must consider an effective separation with low energy consumption to offer a better performance of the system. In our research (Jamsak et al., 2007), we started with a conventional distillation column used in the bioethanol-fuelled SOFC system as illustrated in Figure 6.

A distillation column is commonly recognised as a high energy consumption unit, but the SOFC released a large amount of exothermic heat. Therefore, it is feasible to apply this unit as a bioethanol pretreatment unit. The results from our simulation studies indicated that there were some operating conditions which can run this system under $Q_{net} = 0$. However, the overall electrical efficiency obtained from this system was quite low due to high reboiler heat duty consumption and high amount of heat loss in the condenser. Afterwards, among the promising membrane technologies, pervaporation is considered as a replacement for the former purification unit as shown in Fig 7. By the principle of physical-chemical affinity between the membrane material and species, this unit consumes only heat for vaporizing a preferential substance permeated through the membrane. However, it is noted that a pervaporation depends on a driving force generation device, typically a vacuum pump is used to boost up its separation performance. Therefore, part of the generated electrical energy must be consumed to operate the device.

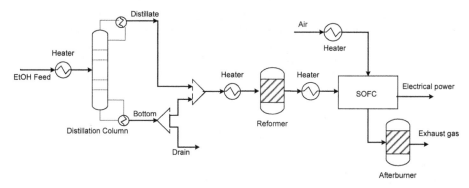

Fig. 6. Process diagram of bioethanol-fuelled SOFC system using a distillation column

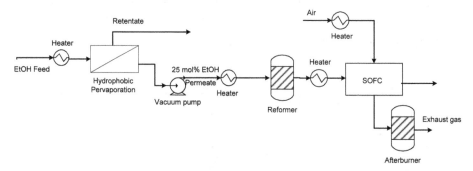

Fig. 7. Process diagram of bioethanol-fuelled SOFC system using a pervaporation

Type of purification process	Energy consumption (MJ/kg ethanol)	Range of concentration (wt%)
Distillation processes Low pressure distillation	11.72	6.4-98
Azeotropic distillation Pentane	10.05	6.4-99.95
Benzene	15.49	6.4-99.95
Diethyl ether	12.56	6.4-99.95
Extractive distillation Gasoline	9.21	6.4-99.95
Ethylene glycol	18.84	6.4-99.95
Extractive distillation with salt Calcium chloride	5.02	7.5-99
Potassium acetate	9.27	60
Non-distillation processes Solvent extraction	6.28	10-98
Pervaporation	4.61	8-99.5

Table 4. Energy consumption for anhydrous ethanol production from various purification processes (Source: Black, 1980; Jaques et al., 1972; Hala, 1969; Barba et al., 1985; Ligero and Ravagnani, 2003)

However, a pervaporation is still regarded as being the lowest energy consumption unit among the other purification units as shown in Table 4. (Reviewed by Kumar et al., 2010) that gives an example of using various purification processes for anhydrous ethanol production. To emphasize their mentioned data, the simulation results from our studies (Choedkiatsakul et al., 2011) showed the performance of bioethanol-fuelled SOFC system in comparison between two pretreatments; using distillation and pervaporation units. On the basis of purification process operated at 348K, Table 5 presents the classification of energy term in each unit for both purification processes. Although a pervaporation consumed an electrical energy within the unit, it offers an overall electrical efficiency (42%) superior to that of distillation column (34%). However, a hydrophobic membrane material used in the pervaporation required a high ethanol separation factor property as illustrated in Figure 8 but it may be unavailable in real membrane materials.

Energy distribution	Purification process configuration	
	Pervaporation	Distillation Column
Heat (MW)		
Bioethanol pretreatment unit	2,301	3,580
Reformer	417	421
Air preheater	22,575	23,892
Afterburner	25,293	27,893
Electrical power (MW)		
Bioethanol pretreatment unit	453	0
Electrical production	4,920	3,701
Net electrical energy	4,467	3,701
Fuel utilization (%)	92	68
Overall electrical efficiency (%)	42	34

Table 5. Performance characteristics in comparison between two different purification units based on Q_{net} = 0, ethanol recovery (R_{EtOH}) = 80%, V = 0.7 V, and T_{SOFC} = 1073 K (Source: Choedkiatsakul et al., 2011)

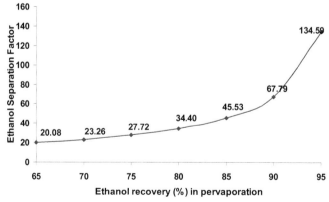

Fig. 8. Effect of ethanol recovery in pervaporation on the required ethanol separation factor of hydrophobic membrane

Consequently, as schematically shown in Figure 9, this problem was solved by having a vapor permeation device installed after a pervaporation (Sukwattanajaroon et al., 2011) to improve ethanol separation performance, an important part of the SOFC system,. The permeate stream of a pervaporation in vapor phase which can be directly fed to a vapor permeator without preheating is a benefit of this technique. From our investigations based on Q_{net} = 0, an available hydrophilic membrane of high water separation factor is a suitable choice to be used in a vapor permeation unit. The performance of SOFC system using this proposed purification process obviously overcomes the other two cases as shown in Figure 10.

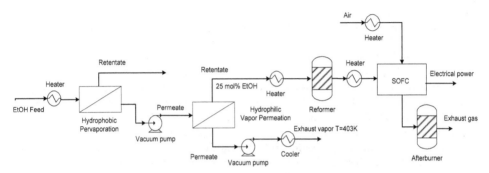

Fig. 9. Process diagram of bioethanol-fuelled SOFC system using a hybrid pervaporation/vapor permeation process

The overall electrical efficiency can be ranked as: Integrated vapour permeation/pervaporation (45.46%) > pervaporation (36.46%) > distillation column (22.53%), respectively.

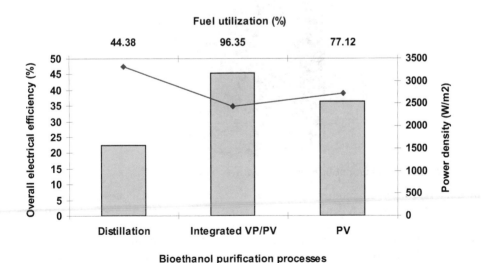

Fig. 10. Performance comparison of SOFC system with various purification processes based on self-sufficient condition (Q_{net} = 0, R_{EtOH} = 75%, V=0.75V, T_{SOFC} = 1073K)

4.2.2 Fuel processing section

Bioethanol was reformed into hydrogen rich gas through a reformer unit which was under the fuel processing section. High yield of desired product and a proper energy management are required for a fuel processor. Structural design of fuel processor is mostly developed to enhance high active surface-to-volume ratio with well-transferred heat. A monolithic reformer is one type of fuel processor design used to increase an active surface area but the compact size of reformer is maintained by designing highly interconnected repeating channels like a honeycomb. The pressure drop along each channel becomes lower. In addition, the monolithic design is resistant to vibration and is stable (Xuan et al., 2009). There is a limit of the temperature control because of its structural design. Nevertheless, heat transfer within the monolithic reformer can be improved by using metallic-typed material as illustrated in Figure 11. Membrane technology is applied to improve the fuel conversion unit which rely on a process integration principle commonly known as membrane reactor. However, Mendes et al., 2010, studied the energy efficiency of the polymer electrolyte membrane fuel cell (PEMFC) system in comparison between a conventional reactor and a membrane reactor operating with ethanol steam reforming. In the case of a conventional reactor, it consists of an ethanol reformer and two water gas shift reactors operating at high and low temperatures, respectively. For a membrane reactor, the multi-tubular module using thin Pd-Ag tubes was employed. The simulation results showed that membrane reactor configuration offers slightly increase of energy efficiency (30%) compared with the conventional reactor (27%) for overall process evaluation. This seems to be impractical because using a membrane-integrated fuel processor requires not only an expensive metal membrane fabrication but also results in short life time due to its low temperature resistance.

Fig. 11. Metallic-made monolithic reactor (Source: Mei et al., 2007)

Internal reforming is another concept of heat allocation techniques similar to process integration can achieve a better efficiency for the SOFC system and also reduce an external reformer cost. A fuel processing section was incorporated with the fuel cell typically placed at an anode side. Heat demand for the endothermic fuel reforming was supplied by the exothermic heat released from the electrochemical reaction of the fuel cell. The operations of internal reforming are classified into two types depending on the level of contact partition between reformer and anode electrode namely; indirect and direct internal reforming as shown in Figure 12.

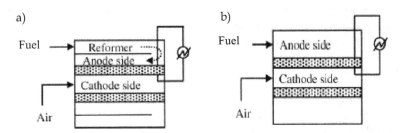

Fig. 12. Type of internal reforming in SOFC: a) Indirect internal reforming, b) Direct internal reforming

Regarding the internal structure of both types of SOFC in Figure 12, direct internal reforming SOFC (DIR-SOFC) can be superior in term of thermal allocation than indirect internal reforming SOFC (IIR-SOFC) owing to its greater contact area of anode electrode. Accordingly, DIR-SOFC can offer a higher overall efficiency. However, by using this operation mode, a carbon formation may occur at on the anode. In our previous work, Assabumrungrat et al., 2004, have investigated the thermodynamic analysis to determine suitable conditions for operating DIR-SOFC fuelled by ethanol to avoid the boundary of carbon formation. From the theoretical calculation results, it is initially suggested that an increase of the H_2O/Ethanol ratio can prevent the carbon formation since adequate water supply leads to a formation of CO_2 rather than CO which is converted to carbon via the Boudard reaction. The oxygen-conducting electrolyte type has lower tendency to form carbon deposition than hydrogen-conducting electrolyte type because the steam product of the first type occurs at the anode side which is the location of fuel processing and thus the additional steam can increase the H_2O/Ethanol ratio in the fuel reforming region.

4.2.3 Heat recovery section

In the SOFC system, after a hydrogen rich gas stream reacted within an SOFC unit under the hydrogen oxidation reaction, an exhaust gas stream containing residual fuel such as H_2, C_2H_5OH, and CO is introduced to combust in the afterburner unit to recover heat for a supply to other heat-demanding units. This brings the system to be more effective heat management and leads to higher system efficiency. There are several methods for recapturing exhaust heat under the frame of combined heat and power (CHP) principle including the use of extra power generation unit (e.g. steam and gas turbines) and heat recovery unit (e.g. recuperator, steam boiler, and heat exchanger network). Selimovic et al., 2002 proposed that networked SOFC stacks incorporated with gas turbine be used to further produce electricity from an exhaust gas combustion stream. It is known that a gas turbine is classified as a low efficiency mechanical power device as well as an entropy lost afterburner. The simulation results showed that the scenario case which allocated fuel utilization portion to the group of afterburner and gas turbine yields higher system efficiency than the scenario case of preferentially allocated fuel utilization portion to the fuel cell. Therefore, it is a good attempt to operate the fuel cell at full performance with high fuel utilization to avoid the step of fuel combustion in the system.

Our previous work (Jamsak et al., 2009) has proposed the MER (maximum energy recovery) under the concept of cogeneration to improve the performance of bioethanol-fuelled SOFC system integrated with distillation column presented in the previous section. Heat transfer arrangement covering useful heat sources i.e. condenser duty, hot water from the bottom of the distillation column, and hot air of cathode recirculation is considered in this study. In the earlier study, system configurations are divided into 5 cases as follows:

a.   Base case (No-HX)
b.   Heat exchanged between the condenser and bioethanol feed stream (CondBio)
c.   Heat exchanged between hot water from the bottom of column and bioethanol inlet stream (HW-Cond)
d.   Heat exchanged between the condenser and air inlet stream (Cond-Air)
e.   Hot air cathode recirculation (CathRec)

All the system configuration studies are illustrated in Figure 13. The basic heat exchanger network was employed in all cases and the results are shown in Table 6.

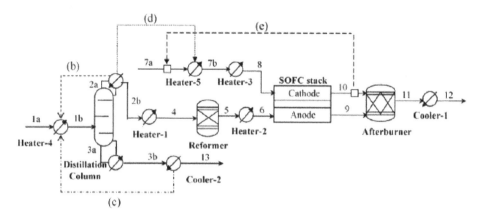

Fig. 13. Process diagram of SOFC system integrated with distillation column: a) No-HX, b) CondBio, c) HW-Cond, d) Cond-Air, and e) CathRec (Source: Jamsak et al., 2009)

Scenario case study	Overall electrical efficiency (%)	CHP efficiency (%)
No-HX	15.79	76.45
CondBio	16.26	78.73
HW-Bio	16.21	78.48
Cond-Air	16.95	81.74
CathRec	21.67	79.87
CondBio-CathRec	22.53	74.71

Table 6. Performance of SOFC system integrated with distillation column with different configurations (Source: Jamsak et al., 2009)

With regard to preheating the bioethanol inlet stream, CondBio can offer both overall electrical efficiency and CHP efficiency higher than those of HW-Bio. Thus, CondBio case is chosen for preheating bioethanol inlet stream. For preheating the air inlet stream, there are two options: Cond-Air and CathRec. Since the condenser has already been used for a bioethanol inlet stream, CathRec has to be selected although its CHP efficiency is slightly less than that of Cond-Air. Afterwards, the CondBio and CathRec are then combined to become a new case: CondBio-CathRec, and its result as shown in Table 6 provides the highest overall electrical efficiency. In addition, this case is further developed by using MER network design. The performance achieved from this design gives 40.8% and 54.3% for overall electrical efficiency and CHP efficiency with the base conditions (25mol%ethanol, ethanol recovery = 80%, operating voltage = 0.7V, fuel utilization = 80%, and operating temperature = 1200K), respectively, compared to the previous SOFC system integrated with distillation column without MER design which gives 33.3% for overall electrical efficiency.

5. Conclusions

This chapter has presented the important use of bioethanol applied as a renewable fuel for producing electricity by Solid Oxide Fuel Cell (SOFC) system. Bioethanol must be upgraded by purifying and reforming into hydrogen rich gas which can be further applied as a clean fuel for direct combustion or electrical power generation by the fuel cell. The later option is chosen as it was realized that bioethanol was worthily utilized in an effective way. The performance development of this system was proposed through our research and the other related scientific literature reviews. Macro level of physical structure design is taken into account for initially guiding a right path for system improvement. Process modification of the system is divided into two main scopes; SOFC and Balance of Plant. The Balance of Plant as a supporting part consists of fuel processing section, bioethanol pre-treatment section, and heat recovery section. All of these are necessary in the concept of process integration and cogeneration to reduce high energy consumption and difficulties within each unit. Bioethanol pretreatment section which is an essential part has been the focus in this chapter. Our evolution of the purification process improvement was proposed. Membrane technology is a promising alternative to be applied in this section and the outcome of SOFC system performance after using this technology is in good agreement with primary mathematical simulation and the criterion of no external energy demand condition. However, an economic assessment and practical experiment in term of investigating working life time should be taken into account for the further study.

6. Acknowledgement

The supports from Thailand Research Fund and National Metal and Materials Technology Center (MTEC), an organization under National Science and Technology Development Agency (NSTDA), are gratefully acknowledged. The authors also would like to thank Dr. John T.H. Pearce for language assistant.

7. References

Assabumrungrat, S.; Pavarajarn, V.; Charojrochkul, S. & Laosiripojana, N. (2004). Thermodynamic analysis for a solid oxide fuel cell with direct internal reforming fuelled by ethanol. *Chemical Engineering Science,* Vol. 59, pp. 6015-6020

Balat, M. (2011). Production of bioethanol from lignocellulosic material via the biochemical pathway: A review. *Energy Conversion and Management,* Vol. 52, pp. 858-875

Barba, D.; Brandini, V. & Di Giacomo, G. (1985). Hyperazeotropic ethanol salted-out by extractive distillation: theoretical evauation and experimental check. *Chemical Engineering Science,* Vol. 40, pp. 2287-2292

Black, C. (1980). Distillation modeling of ethanol recovery and dehydration processes for ethanol and gasohol. *Chemical Engineering Progress,* Vol. 76, pp. 78-85

Bove, R. & Ubertini, S. (2006). Modeling solid oxide fuel cell operation: Approaches, techniques and results. *Journal of Power Sources,* Vol. 159, pp. 543-559

Buchholz, S.E.; Dooley, M.M. & Eveleigh, D.E. (1987). Zymomonas: an alcoholic enigma. *Trends Biotechnology,* Vol. 5, pp. 199-204

Cardona Alzate, C.A. & Sanchez Toro, O.J. (2006). Energy consumption analysis of integrated flowsheets for production of fuel ethanol from lignocellulosic biomass. *Energy,* Vol. 31, pp. 2447-2459

Carlos, A.C. & Oscar, J.S. (2007). Fuel ethanol production: Process design trends and integration opportunities. *Bioresource Technology,* Vol.98, pp. 2415-2457

Chan, S.H.; Khor, K.A. & Xia, Z.T. (2001). A complete polarization model of a solid oxide fuel cell and its sensitivity to the change of cell component thickness. *Journal of Power Sources,* Vol. 93, pp. 130-140.

Choedkiatsakul, I.; Charojrochkul, S.; Kiatkittipong, W.; Wiyaratn, W.; Soottitantawat, A.; Arpornwichanop, A.; Laosiripojana, N. & Assabumrungrat, S. (2011). Performance improvement of bioethanol-fuelled solid oxide fuel cell system by using pervaporation. *International Journal of Hydrogen Energy,* Vol. 36, pp. 5067-5075

Delsman, E.R.; Uju, C.U.; de Croon, M.H.J.M.; Schouten, J.C. & Ptasinski, K.J. (2006). Exergy analysis of an integrated fuel processor and fuel cell (FP-FC) system. *Energy,* Vol. 31, pp. 3300-3309

Deluga, G.A.; Salge, J.R.; Schmidt, L.D. & Verykios, X.E. (2004). Renewable hydrogen from ethanol by autothermal reforming. *Science,* Vol. 303, pp. 993-997

Foyle, T.; Jennings, L. & Mulcahy, P. (2007). Compositional analysis of lignocellulosic materials: evaluation of methods used for sugar analysis of waste paper and straw. *Bioresource Technology,* Vol. 98, pp. 3026-3036

Frison, A. & Memmert, K. (2002). Fed-batch process development for monoclonal antibody production with cellferm-pro. *Genetic Engineering & Biotechnology News,* Vol. 22, pp. 66-67

Gryta, M. (2001). The fermentation process integrated with membrane. *Separation and purification Technology,* Vol. 24, pp. 283-296

Hala, E. (1969). Vapor liquid equilibra in systems of electrolyte components. *Institution of Chemical Engineers Symposium Series,* Vol. 32, pp. 8-16

Jamsak, W.; Assabumrungrat, S.; Douglas, P.L.; Croiset, E.; Laosiripojana, N.; Suwanwarangkul, R. & Charojrochkul, S. (2007). Thermodynamic assessment of solid oxide fuel cell system integrated with bioethanol purification unit. *Journal of Power Sources,* Vol. 174, pp. 191-198

Jamsak, W.; Douglas, P.L.; Croiset, E.; Suwanwarangkul, R.; Laosiripojana, N.; Charojrochkul, S. & Assabumrungrat, S. (2009). Design of a thermally integrated bioethanol-fuelled solid oxide fuel cell system integrated with a distillation column. *Journal of Power Sources,* Vol. 187, pp. 190-203

Jaques, D. & Furter, W.F. (1972). Salt effects in vapour-liquid equilibrium: testing the thermodynamic consistency of ethanol-water saturated with inorganic salt. *AIChE Journal,* Vol. 18, pp. 343-346

Kakac, S.; Pramuanjaroenkij, A. & Zhou, X.Y. (2007). A review of numerical modeling of solid oxide fuel cells. *International Journal of Hydrogen Energy,* Vol. 32, pp. 761-786.

Kumar, S.; Singh, N. & Prasad, R. (2010). Anhydrous ethanol: A renewable source of energy. *Renewable and Sustainable Energy Reviews*, Vol. 14, pp. 1830-1844

Lal, R. (2008). Crop residues as soil amendments and feedstock for bioethanol production. *Waste Management*, Vol.28, pp. 747-758

Ligero, E.I. & Ravanani, T.M.K. (2003). Dehydration of ethanol with salt extractive distillation-a comparative analysis between processes with salt recovery. *Chemical Engineering Process*, Vol. 42, pp. 543-552

Liguras, D.K.; Kondarides, D.I. & Verykios, X.E. (2003). Production of hydrogen for fuel cells by steam reforming of ethanol over supported noble metal catalysts. *Applied Catalysis B: Environmental*, Vol. 43, pp. 345-354

Magnusson, H. (2006). Process simulation in Aspen Plus of an integrated ethanol and CHP plant. *Master Thesis in Energy Engineering*, Department of Applied Physics and Electronics, Umea University, Sweden.

Mei, H.; Li, C.; Ji, S. & Liu, H. (2007). Modeling of a metal monolith catalytic reactor for methane steam reforming-combustion coupling. *Chemical Engineering Science*, Vol. 62, pp. 4294-4303.

Mendes, D.; Tosti, S.; Borgognoni, F.; Mendes, A. & Madeira, L.M. (2010). Integrated analysis of a membrane-based process for hydrogen production from ethanol steam reforming. *Catalysis Today*, Vol. 156, pp. 107-117

Minh, N.Q. (1993). Ceramic Fuel Cells. *Journal of the American Ceramic Society*, Vol. 76, pp. 563-588

Nagel, F.P.; Schildhauer, T.J.; Biollaz, S.M.A. & Wokaun, A. (2008). Performance comparison of planar, tubular and Delta8 solid oxide fuel cells using a generalized finite volume model. *Journal of Power Sources*, Vol. 184, pp. 143-164

Ni, M.; Leung, D.Y.C. and Leung, M.K.H. (2007). A review on reforming bio-ethanol for hydrogen production. *International Journal of Hydrogen Energy*, Vol. 32, pp. 3238-3247.

Paul, N. & Kemnitz, D. (2006). Biofuels – Plants, Raw Materials, Products. – Fachagentur Nachwachsende Rohstoffe e.V. (FNR), *WPR Communication*, Berlin, pp. 43

Petruzzi, L.; Cocchi, S. & Fineschi, F. (2003). A global thermo-electrochemical model for SOFC systems design and engineering. Dependence of polarization in anode-supported solid oxide fuel cells on various cell parameters. *Journal of Power Sources*, Vol. 118, pp. 96-107

Piroonlerkgul, P.; Assabumrungrat, S.; Laosiripojana, N. & Adesina, A.A. (2008). Selection of appropriate fuel processor for biogas-fuelled SOFC system. *Chemical Engineering Journal*, Vol. 140, pp. 341-351

Pramuanjaroenkij, A.; Kakac, S. & Zhou, X.Y. (2008). Mathematical analysis of planar solid oxide fuel cells. *International Journal of Hydrogen Energy*, Vol. 33, pp. 2547-2565

Roger, P.L.; Lee, K.J. & Tribe, D.E. (1980). High productivity ethanol fermentations with Zymomonas mobilis. *Process Biochemistry*, pp. 7-11

Selimovic, A. & Palsson, J. (2002). Networked solid oxide fuel cell stacks combined with a gas turbine cycle. *Journal of Power Sources*, Vol. 106, pp. 76-82

Shell, D.J.; Riley, C.J.; Dowe, N.; Farmer, J.; Ibson, K.N.; Ruth, M.F.; Toon, S.T. & Lumpkin, R.E. (2004). A bioethanol process development unit: initial operating experiences and results with a corn fiber feedstock. *Bioresource Technology*, Vol. 91, pp. 179-188

Sukwattanajaroon, V.; Charojrochkul, S.; Kiatkittipong, W.; Arpornwichanop, A. & Assabumrungrat, S. (2011). Performance of membrane-assisted solid oxide fuel cell system fuelled by bioethanol. *Engineering Journal,* Vol. 15, pp. 53-66

Tao, G.; Armstrong, T. & Virkar, A. (2005). Intermediate temperature solid oxide fuel cell (IT-SOFC) research and development activities at MSRI. *In: Nineteenth annual ACERC&ICES conference,* Utah

U.S. Department of Energy. Energy Efficiency & Renewable Energy, http://www1.eere. energy.gov/biomass/abcs_biofuels.html

U.S. Environmental Protection Agency, http://www.epa.gov/methane/.

Vourliotakis, G.; Skevis, G. & Founti, M.A. (2009). Detailed kinetic modeling of non-catalytic ethanol reforming for SOFC applications. *International journal of hydrogen energy,* Vol. 34, pp. 7626-7637

Wilkins, M.R.; Widmer, W.W. & Grohmann, K. (2007). Simultaneous saccharification and fermentation of citrus peel waste by Saccharomyces cerevisiae to produce ethanol. *Process Biochemistry,* Vol. 42, pp. 1614-1619

Xuan, J.; Leung, M.K.H.; Leung, D.Y.C. & Ni, M. (2009). A review of biomass-derived fuel processors for fuel cell systems. *Renewable and Sustainable Energy Reviews,* Vol. 13, pp. 1301-1313

Zhao, F. & Virkar, A.V. (2005). Dependence of polarization in anode-supported solid oxide fuel cells on various cell parameters. *Journal of Power Sources,* Vol. 141, pp. 79-95

9

Combustion and Emissions Characteristics of Biodiesel Fuels

Chaouki Ghenai
Ocean and Mechanical Engineering Department, Florida Atlantic University
United States of America

1. Introduction

The production and use of biodiesel as an alternative diesel fuel in compression-ignition engines and boilers has increased significantly in the recent years. Biodiesel is considered to be an immediate alternative energy, providing a solution to help decrease the effects of harmful global green house gases, why decreasing the dependency of fossil fuels (Demirbas, 2008, Gärtner and Reinhardt, 2003). Biodiesel is derived from plant oils, animal fats and recycled cooking oils (Biodiesel Handling and Use Guide, 2009). Bio-Diesel is a renewable fuel produced by a chemical reaction of alcohol and vegetable or animal oils, fats, or greases. Bio-Diesel offers a safer and cleaner alternative to petroleum Diesel. Biodiesel is renewable fuel, its is energy efficicient, it can be used as a 20% blend in most diesel equipment with no or only minor modifications, can reduce global warming gas emissions, it is nontoxic, biodegradable, and suitable for sensitive environments (Biodiesel Handling and Use Guide, 2009).

Biodiesel is produced when vegetable oil or animal fat is chemically reacted with alcohol (methanol or ethanol) in the presence of catalyst such as sodium or potassium hydroxide (Van Gerpen, 2004). Glycerin is produced as a co-product. Biodiesel fuel is produced from oil feedstock such as soybean oil, corn oil, canola oil, cottonseed oil, mustard oil, palm oil, restaurant waste oils such as frying oils, animal fats such as beef tallow or lard, trap grease (from restaurant grease traps), float grease (from waste water treatment plants - Van Gerpen, 2004). The oil or animal fat can be converted to methyl or ethylesters (biodiesel) directly, using a base reaction (catalyze) to accelerate the transesterification reaction. The most common method of production of biodiesel is by mixing the vegetable oil with methanol in the presence of sodium hydroxide.

The reaction produces methyl esters (Biodiesel) and glycerin (by product). Biodiesel can be used in its pure form B100, which requires some modification to the engine, to prevent any decomposition of plastic parts. Because the level of special care needed is high, the National Renewable Energy Laboratory (NREL) and the U.S. Department of Energy (DOE) do not recommend the use of high-level biodiesel blends. When human exposure to diesel particulate matter (PM) is elevated, additional attention to equipment and fuel handling is needed (Biodiesel Handling and Use Guide, 2009). More commonly biodiesel is run as a blend, such as B5, B10, and B20 (Example: B20 is 20% of biodiesel blended with 80% of petroleum diesel). No modification of engine is needed if Biodiesel fuel blends are used. At concentrations of up to 5 vol % (B5) in conventional diesel fuel, the mixture will meet the ASTM D975 diesel fuel specification and can be used in any application as if it was pure petroleum diesel; for home heating oil, B5 will meet the D396 home heating oil specification (Biodiesel Handling and Use Guide, 2009). At concentrations of 6% to 20%, biodiesel blends can be used in many applications that use diesel fuel with minor or no modifications to the equipment. B20 is the most commonly used biodiesel blend in the United States because it provides a good balance between material compatibility, cold weather operability, performance, emission benefits, and costs (Biodiesel Handling and Use Guide, 2009). Equipment that can use B20 includes compression-ignition (CI) engines, fuel oil and heating oil boilers, and turbines. The analysis, fuel quality, and production monitoring of biodiesel have been discussed in more details in previous studies (Knothe, 2005, Mittelbach and Remschmidt, 2004, Knothe, 2001, Mittelbach, 1996 and Komers et al., 1998).

2. Biodiesel basics

Biodiesel is produced from plant oils (soybean oil, cotton seed oil, canola oil), recycled cooking greases or oils (e.g., yellow grease), or animal fats (beef tallow, pork lard). Biodiesel is the result of a chemical reaction process on oils or fats called transesterification. A simple diagram of the transesterification process is shown in Figure 1. Vegetable oil, animal fats or waste oil react with alcohol in the presence of catalysts to form Biodiesel and Glycerin. Glycerin is a co-product of the biodiesel process. There are three distinct types of transesterification process: (1) Base catalyzed transesterification of the oil, (2) Direct acid catalyzed transesterification of the oil, and (3) conversion of the oil to its fatty acids and then to biodiesel. For the direct acid catalyzed transesterification of the oil, the transesterification process is catalyzed by bronsted acids, preferably by sulfonic and sulfuric acids. These catalysts give very high yields in alkylesters, but the reactions are slow, requiring, typically, temperatures above 100 °C and more than 3 h to reach complete conversion. The base-catalyzed transesterification of vegetable oils proceeds faster than the acid-catalyzed reaction and the alkaline catalysts are less corrosives than acidic compounds. From these three basic route to biodiesel production the base catalyzed transesterification of the oil present many advantages. First the base catalyzed transesterification has high conversion efficiency. With this process around 98 percent of all the reactants will effectively mix to produce biodiesel. This chemical reaction is also efficient at low temperature and low pressure. This is great advantages because it means that no heavy pressure unit or heating unit would be required. Furthermore this transesterification lead to a direct conversion and does not need any intermediate compounds to achieved biodiesel production.

2.1 Reactants

Two reactants are present during the chemical reaction. The first one is the triglycerides from the vegetable oil and the second is the alcohol. Triglycerides can have different alkyl groups as biodiesel can be made out of different kind of straight vegetable oil or waste vegetable oil. Two different kind of alcohol can be used: methanol or ethanol. These two alcohols are used in particular for this chemical reaction because there is very little space between triglycerides atoms for the alcohol to react with.

2.2 Products

From this chemical reaction there are two main products: Glycerol also named glycerin and methyl or ethyl ester which is Biodiesel. Glycerin is denser than biodiesel and would consequently stay under it.
The equation below will resume the reaction:

$$\text{Vegetable oil + alcohol} \longrightarrow \text{glycerin + Biodiesel} \tag{1}$$

If we consider the concentrations:

$$80\% \text{ Vegetable Oil} + 20\% \text{ Alcohol} \longrightarrow 90\% \text{ Biodiesel} + 10\% \text{ Glycerin} \tag{2}$$

2.3 Catalyst

The catalyst if added in the good proportion will neutralize the free fatty acids (FFA) from the waste vegetable oil and also accelerates the reaction. For this type of transesterification two different catalysts are usually used: Sodium Hydroxide NaOH or Potassium Hydroxide KOH. The catalyst is used to accelerate the chemical reaction. The quantity of catalyst needed depend on the amount of free fatty acids in the waste oil and will be determine by the titration of the oil (process used for the determination of the right amount of catalyst that will be use for the chemical process).

2.4 Transesterification of vegetable oils

In the transesterification of vegetable oils (see Fig.2), a triglyceride reacts with an alcohol in the presence of a strong acid or base, producing a mixture of fatty acids alkyl esters and glycerol (Wright et al., 1944 and Freedman, 1996). The stoichiometric reaction requires 1 mol of a triglyceride and 3 mol of the alcohol. However, an excess of the alcohol is used to increase the yields of the alkyl esters and to allow its phase separation from the glycerol formed. Several aspects, including the type of catalyst (alkaline or acid), alcohol/vegetable oil molar ratio, temperature, purity of the reactants (mainly water content) and free fatty acid content have an influence on the course of the transesterification (Schuchardta, 1998) based on the type of catalyst used.

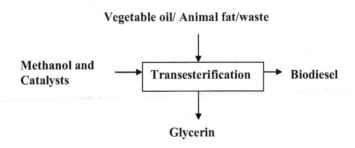

Fig. 1. Basic transesterification process

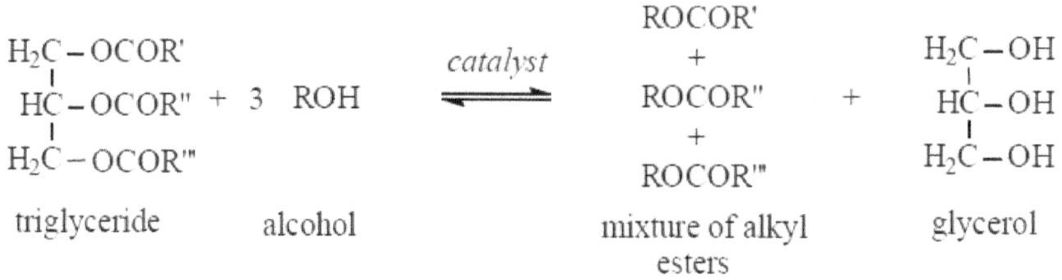

$$\begin{array}{l} H_2C-OCOR' \\ | \\ HC-OCOR'' \\ | \\ H_2C-OCOR''' \end{array} + 3 \ ROH \ \underset{}{\overset{catalyst}{\rightleftharpoons}} \ \begin{array}{c} ROCOR' \\ + \\ ROCOR'' \\ + \\ ROCOR''' \end{array} + \begin{array}{l} H_2C-OH \\ | \\ HC-OH \\ | \\ H_2C-OH \end{array}$$

triglyceride   alcohol   mixture of alkyl   glycerol
                                                              esters

Fig. 2. Transesterification of Vegetable Oils

The principal objective of this study is to produce quality biodiesel fuels using different oil feed stocks and test the performance and emissions of Diesel engine using different biodiesel fuel blends (B5, B10, B15 and B20). The combustion performance (torque and engine horsepoqer) and emissions (CO, CO2, HC's, and NOX) from the Diesel engine using a Bio-Diesel fuel blends and the conventional petroleum Diesel fuel will be compared.

3. Biodiesel production and engine testing

3.1 Biodiesel production

The biodiesel fuel was produced through the base catalyst transterification process. This provides one of the easiest and more efficient ways to produce the fuel, yielding a return of close to 98%. A biodiesel processor (see Figure 3) with a production capacity of 40 gallons/day was used to produce the biodiesel. The processor contains two tanks: (1) a small tanks for mixing the base catalyst and methanol to create the methoxide required for the reaction, and (2) bigger tank (40 gallons) for mixing the oil with methoxide. The processor contains a small 120V pump which transfers the methoxide into the base oil for mixing. The processor is also equipped with heated coil (in the main reaction tank) to speed up the time required to produce the biodiesel fuel and the glycerin.

Four different oil feedstocks (see Fig .4) were tested in thi study: (1) Waste vegetable oil from Coyote Jacks, (2) Waste vegetable oil from JC Alexander, (3) straight or non used peanit oil, and (4) straight or non used coconnut oil. It is noted that Coyote Jacks and JC Alexander are two dinning facilities inside and outside the University. Waste and straight (non used) oils were testd in this to see the effcet of the quality of oil feedstocks on the produced biodiesl fuesl. For the waste vegetbale oil, the waste oil is first heated up to 120°F to evaporate any water present and filtered to eliminate any solid particles in the oil. For all the oil feedstocks use in this study, a titration was performed to determine the exact amount of base catalyst needed for the reaction based on the amount of biodiesel fuel to be produced.

The mixing of the base oil (See Figure 3) and methoxide (methol and cataltyst) will help to break down all the fatty amino acids present in the base oil. The mixing process takes up to 1 hour. After that, we let the mixture to settle down for 7 to 8 hours. This will help to separate the Biodiesel fuel from Glycerin as shown in Fig. 5 (the Biodiesel is on the top and Glycerin in the bottom). After the separation of Glycerin from the Biodiesel, the fuel is washed with water and dried to remove any excess methanol and glycerin present in the final product (B100 Biodiesel fuel). The quality of the final biodiesel product is tested in the laboratory and the correponsding biodiesel fuel blends (B5, B10, B15, B20) are prepared for the engine testing. It is noted that B20 for example is 20% biodiesel and 80% petroleum diesel fuel.

Fig. 3. Biodiesel Processor with a capacity of 40 gallons/day

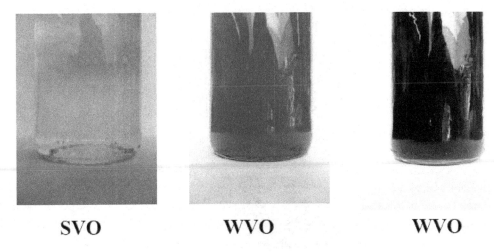

Fig. 4. Straight and waste vegetable oil before transesterification

Fig. 5. Biodiesel and Glycerin

3.2 Diesel engine

The Diesel engine used for this study is an air cooled, 4-cycles, single cylinder and 8 horse power Robin Engine. The Diesel engine is compact, lightweight and designed for generators, air compressor, water pumps, pressure washer and light construction machines (see Figure 6). The engine has a fuel tank of a capacity of approximately 14.5 liters (3.83 gallons), and a maximum horsepower rating of 8 HP at 3600 rotation per minute. The piston displacement is 348 cubic centimeters. The bore and stroke are respectively 82 and 78 mm. The engine combustion system has a direct injection. The engine dimensions are 435x370x478 mm and the engine weight is 46 Kg. A fuel flow meter is connected to the engine (see Figure 7) to record in real time the fuel consumption during engine testing.

Fig. 6. 8 Horse Power Diesel Engine

Fig. 7. Fuel flow meter

3.3 Diesel engine torque and horse power measurements

A small engine dynamometer (See Figure 8) was used to test the performance (torque and horse power) of the Diesel engine. The dynamometer is a water brake load system. The pump pressurizes the water which then goes into the water brake connected to the engine shaft. Part of the energy produced by the engine goes to the exhaust while the rest of it is released as calorific calories to the water that passes thru the water brake. For this reason the water has to be cooled down, as it will over heat over time and loose efficiency. The dynamometer is equipped with sensors, data acquisition system and software to measure the Torque and engine RPM (rotation per minutes). The measured values of the torque and RPM are used to determine the horsepower:

$$\text{Horsepower} = (\text{RPM} * \text{Torque}) / 5252 \qquad (3)$$

Fig. 8. Engine Dynamomter

3.4 Emissions characterization - Engine exhaust gas measurements

A gas analyzer DYNOmite EMS (See Figure 9) was used in this study to measure the emissions from the engine. The real time gas analyzer is used to measure the concentrations of oxygen (O_2), carbon monoxide (CO), carbon dioxide (CO_2), hydrocarbons (HC) and nitrogen oxides NOx. The gas emissions from the engine are recorded simultaneously during engine dynamometer testing using the data acquisition system. The torque; RPM; horse power; O_2, CO, CO_2, $HC's$ and NOx concentrations are measured simultaneously.

Fig. 9. Gas Analyszer and emission probe

3.5 Data acquisition system

The dynamometer data acquisition system (see Figure 10) is used to record the engine performance (Torque, RPM, and Horse power), fuel consumption and the exhaust gas emissions (CO, CO_2, NOx, and HCs). All these data are recorded simultaneously during engine testing.

4. Results and discussions

4.1 Combustion and emission characteristics of petroleum diesel fuels

Figures 11 and 12 show the engine performance and emissions using conventional petroleum Diesel fuel at different engine speed or rotation per minutes (RPM). The engine starts running at low RPM (~ 1230 RPM) for six seconds after that the RPM of the engine was increased until it reached the maximum value of 3650 RPM. The engine performances (engine horse power and torque) and emissions (CO_2, CO, HC and NOX) were recorded for about 18 seconds during this change of the engine RPM. Figure 11 shows that horse power increases from 2.3 HP at low RPM to about 8 HP at high RPM and the torque increases from 10 ft lb at low RPM to about 12 ft lb at high RPM. Figure 12 shows the variation of the engine emissions (CO_2, CO, NOX, HCs) with the engine speed or RPM.

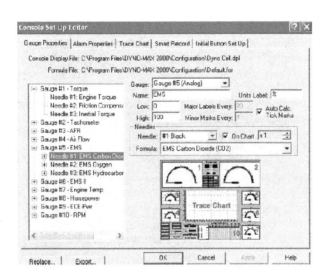

Fig. 10. Data acquisition system

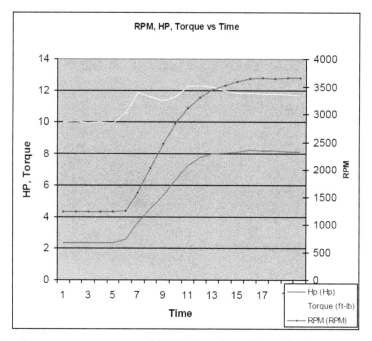

Fig. 11. Engine horse power, torque, and RPM - Petroleum Diesel fuel

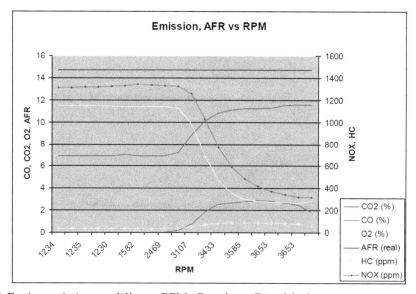

Fig. 12. Engine emissions at different RPM - Petroleum Diesel fuel

4.2 Combustion and emission characteristics of biodiesel fuel blends

Four different oil feed stocks were used in this study to produce pure bio-diesel fuel. The Biodiesel fuel produced from each oil feed stocks was converted into fuel blends of B5, B10, B15, and B20. The Biodiesel fuel blends are obtained by mixing biodiesel fuel with petroleum Diesel fuel (ex: B5: 5% Biodiesel and 95% Diesel fuel by volume). Two of the oil feed stocks are waste cooking oil collected from two different restaurants. The first waste cooking oil was collected from Coyote Jacks, a restaurant located inside the premises of Florida Atlantic University. This restaurant is more focused on deep fried fast food, and the oil is used over and over for up to 3 weeks before it is changed. The second waste cooking oil comes from J.C. Alexander restaurant located outside Florida Atlantic University. This restaurant uses its oil for shorter periods of time, and is mainly used to sauté food instead of deep frying. The reason for using two different waste cooking oils from two different places is to see if the quality of the oil feed stocks will affect the quality of the biodiesel produced. The third and fourth oils tested in this study were Peanut and Coconut oils (not used oils). The peanut and coconut oils have not been used for cooking prior to being converted into bio-diesel fuel. The procedure for each test was: (1) the engine was allowed to run for 2 minutes to allow the engine to clear any residue from previous tests with different fuel blends, and warm up the engine and catalyst, (2) after the 2 minutes warm up, the engine was run at high RPM (3550 rpm) for 5 minutes, (3) drop the engine speed to medium RPM (2400 rpm) and run the engine for 5 minutes, and (4) finally set the engine to low RPM (1250 rpm) and run for 5 minutes. The sampling rate was set to 1 Hz. This will produce a total of 300 data points for each RPM. For each Biodiesel fuel blend and for each RPM, the measurement was run at least three times to check the consistency of the data produced. The data produced from these tests includes: Torque (lbs-feet), Horse Power (HP), Carbon Dioxide CO_2 (%), Carbon Monoxide CO (%), Oxygen O_2 (%), Hydro Carbon HC (ppm) and Nitric Oxide NOx (ppm). The torque and horsepower readings where measured directly from the dynamometer, while the emissions where obtained from the gas analyzer. The mean value (over 5 minutes) of the diesel engine tests for low, medium and high RPM using petroleum diesel (benchmark) are shown in Table 1. It is noted that the horse power, CO_2 and CO emissions increase and the HC and NOx emissions decrease when the speed (rpm) of the engine increases. The baseline data with the petroleum Diesel fuel will be compared to those obtained with Biodiesel fuel blends using different type of oil feed stocks. Typical results of the Diesel engine tests at high RPM using biodiesel fuel blends (B5, B10, B15 and B20) produced from waste vegetable oil coyote Jacks are shown in Table 2. The results show a small change of the engine horse power, torque, CO_2 and NOx emissions when the amount of biodiesel blended with petroleum Diesel increases from 5% to 20%. The CO emissions dropped from 0.27% to 0.25% and the HC emissions from 14.06 ppm to 12.14 ppm.

RPM	HP	Torque Lbs-feet	CO2 %	CO %	O2 %	HC ppm	NOx ppm
1250	2.34	9.84	7.28	0.02	12.10	33.74	1157
2400	4.50	9.86	7.58	0.02	12.30	30.77	617
3550	7.97	11.80	11.45	0.28	3.98	14.50	398

Table 1. Mean value of the Diesel engine tests using petroleum Diesel

	B5	B10	B15	B20
HP	7.89	7.82	7.80	7.88
Torque, Lbs-feet	11.71	11.66	11.58	11.49
CO2 %	11.68	11.70	11.60	11.43
CO %	0.27	0.26	0.259	0.25
O2 %	3.55	3.50	3.49	3.47
HC ppm	14.06	13.18	12.82	12.14
NOx ppm	398	402	405	406

Table 2. Mean value of the Diesel engine tests at high RPM (3550) using Biodiesel Fuel Blends produced from waste vegetable oil Coyote Jacks

The results obtained in this study with the three other oil feed stocks (waste cooking oil JC Alexander, peanut and coconut oils) show the same trends. Figure 13 shows the percentage difference of the data obtained with the Biodiesel fuel blend B20 from the four oil feed stocks and the petroleum diesel fuel. The results show a net decrease of the Hydrocarbons HC and CO emissions for the B20 biodiesel fuel blend compared to the petroleum diesel fuel for all the four oil feed stocks. Biodiesel fuel contains fewer hydrocarbons than those present in petroleum diesel. It is natural to expect a decrease in HC emissions as the blends increase. The Hydrocarbons emissions for the B20 decreases by about 20% compared to the petroleum Diesel fuel. The carbon monoxide emissions decrease by about 12 % for the B20. Although the hydrocarbon HC and CO emissions decreased drastically, a small change (< 2%) of the engine power HP and CO2 emissions are reported for the B20 compared to the petroleum Diesel fuel. The NOx emissions also increase by about 2% for the B20. This is due probably to an increase of thermal NOx because of the combustion temperature increase for the B20.

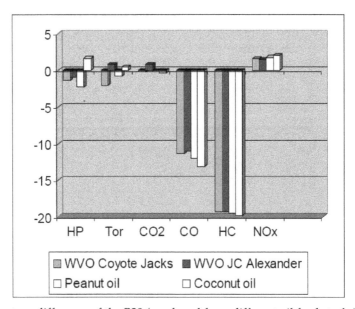

Fig. 13. Percentage difference of the B20 (produced from different oil feed stocks) with Respect to Petroleum Diesel

Figure 14 shows the overall effect by averaging all the four diverse oils and normalizing them over the petroleum diesel (baseline data). It is shown clearly the benefits of Biodiesel fuel blends compared to the petroleum diesel fuel. The HC and CO emissions decrease drastically by increasing the amount of biodiesel blended with Diesel fuel, while the power of the engine is kept almost the same and the NOx emissions increase by not more than 2%.

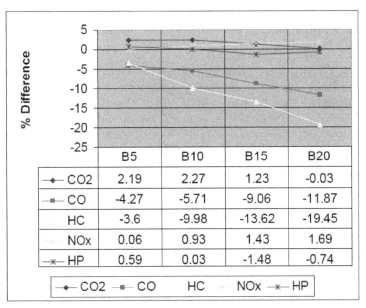

	B5	B10	B15	B20
CO2	2.19	2.27	1.23	-0.03
CO	-4.27	-5.71	-9.06	-11.87
HC	-3.6	-9.98	-13.62	-19.45
NOx	0.06	0.93	1.43	1.69
HP	0.59	0.03	-1.48	-0.74

Fig. 14. Overall percentage differences of biodiesel fuel blends with respect to petroleum Diesel

4. Conclusions

Biodiesel fuel was produced from four different oil feed stocks: waste cooking oil Coyote Jacks, waste cooking oil JC Alexander, and not used peanut and coconut oils. Biodiesel fuel blends (B5, B10, B15 and B20) for each oil feed stock were prepared by mixing the biodiesel fuel with Petroleum Diesel. The biodiesel fuel blends were tested on the Diesel Engine and the results were compared to those obtained with Petroleum Diesel Fuel. The results obtained with the four oil feed stocks show the same trends. The hydrocarbons HC and CO emissions decreased by increasing the amount of biodiesel blended with Petroleum Diesel. The HC and CO emissions decreased respectively by 20 % and 12 % for the B20 compared to the Petroleum Diesel while the NOx emissions increased by 2% and the change of the engine power was negligible (<2%).

5. References

Demirbas, Ayhan. "Biodiesel: A Realistic Fuel Alternative for Diesel Engines", Springer, New York, 2008

Gärtner, Sven O. and Reinhardt, Guido A., Ph.D. "Life Cycle Assessment of Biodiesel: Update and New Aspects", Heidelberg, May 2003

Biodiesel Handling and Use Guide Fourth Edition, Nationa Renewable Energy Laboratory, Onnovation of Our Energy Future, NREL/TP-540-43672 Revised January 2009

Van Gerpen, J., Shanks, b., Pruszko, r., Clements, d., and Knothe, G., Biodiesel Analytical Methods, National Renewable Energy Laboratory, NREL/SR- 510-36240, 2004

Knothe, G., J. Van Gerpen, and J. Krahl (eds.), The Biodiesel Handbook, AOCS Press, Champaign, Illinois, 2005, 302 pp.

Mittelbach, M., and C. Remschmidt, Biodiesel — The Comprehensive Handbook, published by M. Mittelbach, Karl-Franzens- Universität Graz, Graz, Austria, 2004.

Knothe, G., Analytical Methods Used in the Production and Fuel Quality Assessment of Biodiesel, Trans. ASAE 44:193–200 (2001).

Mittelbach, M., Diesel Fuel Derived from Vegetable Oils, VI: Specifications and Quality Control of Biodiesel, Bioresour. Technol. 56:7–11 (1996).

Komers, K., R. Stloukal, J. Machek, F. Skopal, and A. Komersová, Biodiesel Fuel from Rapeseed Oil, Methanol, and KOH. Analytical Methods in Research and Production, Fett/Lipid 100:507–512 (1998).

Wright, H.J.; Segur, J.B.; Clark, H.V.; Coburn, S.K.; Langdon, E.E.; DuPuis, R.N., Oil & Soap, 21, pp. 145-148, 1944.

Freedman, B.; Butterfield, R.O.; Pryde, E.H. J. Am. Oil Chem. Soc. 1986, 63, 1375

Schuchardta Ulf, Ricardo Serchelia, and Rogério Matheus Vargas, Transesterification of Vegetable Oils: a Review, J. Braz. Chem. Soc., Vol. 9, No. 1, 199-210, 1998.

10

Data Acquisition in Photovoltaic Systems

Valentin Dogaru Ulieru[1], Costin Cepisca[2],
Horia Andrei[1] and Traian Ivanovici[1]
[1]Valahia University of Targoviste
[2]University Politehnica of Bucharest
Romania

1. Introduction

During the last decades, human lifestyle and economic growth has had a profound effect on the energetic sector considerably changing the perspective on the energy issue (Ambros et al., 2004). The increasing energy demand and variable oil price, insecure energy resources and carbon dioxide emission have made us aware of the fact that energy is indeed a limited product (Awerbuch, 2002).

Regarding energy resources, the International Energy Agency estimates that oil resources will be over in 40 years, natural gas resources in 60 years and coil resources in 200 years. Renewable energy and coal are the fastest growing energy sources, with consumption increasing by 2.1 percent per year and 2.0 percent, respectively. A significant number of studies and scenarios have investigated the contribution of renewable energy to satisfy global needs in energy, indicating that during the first half of XXI century its contribution will increase from 20 to 50%.

Estimating the exploitable technical potential of renewable energy in Romania, it observed there is a high potential of our country, for the usage this type of renewable energy, and Romania's strategy in this area provides for 2012 a energy production of: 1860 MWh from photovoltaic (PV) sources, 314,000 MWh wind sources; 18,200,000 MWh hydroelectric sources and 1,134,000 MWh biomass. Total of 19,650,000 MWh should represent 30% of country`s electricity consumption (Vasile, 2009).

PV systems produce power in proportion to the intensity of sunlight striking the solar array surface. The intensity of light on a surface varies throughout a day, as well as day to day, so the actual output of a solar power system can vary substantialy. There are other factors that affect the output of a solar power system. These factors need to be understood so that the customer has realistic expectations of overall system output and economic benefits under variable environmental conditions over time. From this perspective, the development of photovoltaic systems is closely linked to development of measurement and monitoring techniques, built-in data acquisition systems.

Data acquisition systems (DAQ) can measure and store data collected from hundreds of channels simultaneously. The majority of systems contain from eight to 32 channels, typically in multiples of eight. An ideal data acquisition system uses a single ADC for each measurement channel. In this way, all data are captured in parallel and events in each channel can be compared in real time. But using a multiplexer that switches among the inputs of multiple channels and drives a single ADC can substantially reduce the cost of a system (Szekely, 1997).

Specialized data acquisition systems for PV installations require a study of sample rates and an optimal configuration of the measuring chain. This chapter brings informations regarding the structure of data acquisition systems used in the monitoring of photovoltaic installations. It shows the operating principles of building blocks and the operation is performed by simulations using LabVIEW™ - Laboratory Virtual Instrumentation Engineering Workbench.

An important part of the presentation is dedicated to current measurement and data acquisition systems dedicated for monitoring PV systems. Applied solutions and experimental results are discussed in terms of accuracy and optimization needs of the operation.

2. PV system and data acquisition

The chapter presents the authors activity and results regarding the operation of a PV system and aspects on monitoring the electric energy supplied by a PV system built in University "Valahia" of Targoviste, Romania.

a) North view   b) South view

Fig. 1. PV system

Part of the ICOP DEMO 4080-90 European research program, this PV system has been realized by the staff of the Electrical Engineering Faculty, Targoviste, Romania (Andrei et al., 2007).

The PV system integrated into the roof of the building has been designed using 66 Optisol SFM 72Bx glass roof integrated multi crystalline Si modules produced by Pilkington Solar and 24 ST40 thin film modules produced by Siemens which generates a total amount of 10 kWp. Position of panels on the southern front is shown in Figure 2. These modules can be serial or parallel connected. The dc voltage produced by the PV system is converted by the Sunny Boy inverters (SWR 700, SWR 1100, SWR 2000 and SWR 2500) and supplied directly into the public electricity system – Figure 3.

The use of a controller to monitor the operating parameters ensures the sine-wave form of the voltage and current, with a low amount of harmonics. The control of operations serves to totally automated functioning and to adjustement of the MPP (maximum power point).

The connection diagram of the PV system has been designed after a series of shading effects analyse and buiding placement restrictions (Dogaru Ulieru et al., 2009).

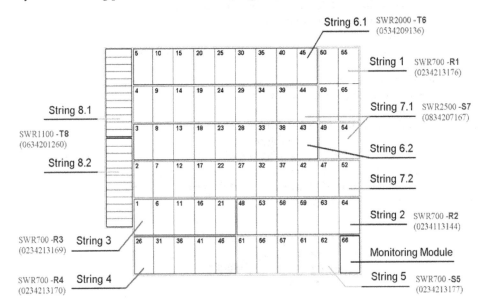

Fig. 2. Panel position

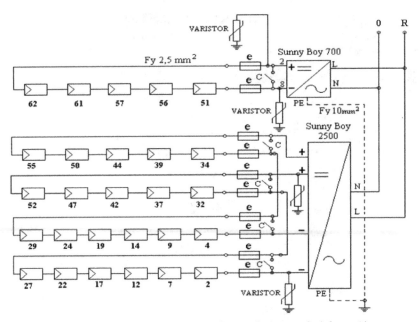

Fig. 3. Electric energy producing system using photovoltaic panels (phase A)

PV system monitoring was done by two methods:

-   Sunny Data Control equipment – Figure 4 - automatically reads the selected measuring data from the memory of Sunny Boy Control and stores it on the PC. Sampling interval has been preset to 10 minutes (day and night).
-   The data acquisition system - represented by an association between the hardware equipment (AT-MIO 16XE50 acquisition board, a signal conditioning device) and the application software (LabVIEW™) which implements the required functions, playing the part of an interface between the human operator and the measurement system .

To ensure the accuracy of the measurement, the operating parameters of the PV system and the configuration of the acquisition system are taken into account and have imposed the signal conditioning and the setting of the signal source, of the field and of the channels. Analog inputs can be differently configured, with a voltage level of ±2.5V, ±5V, ±10V (bipolar/single polar) which can be selected through the configuration program of the acquisition board. The block diagram of the acquisition system is presented in Figure 5. Figure 6 shows the LabVIEW™ block diagram of the virtual instrument attached to the data acquisition system (Cepisca et al., 2004).

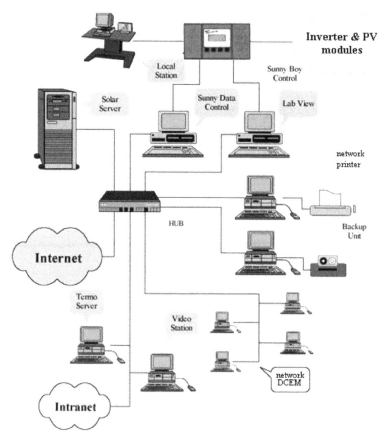

Fig. 4. Monitoring system Sunny

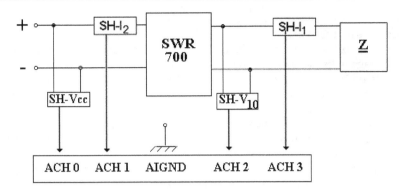

Fig. 5. DAQ Block diagram

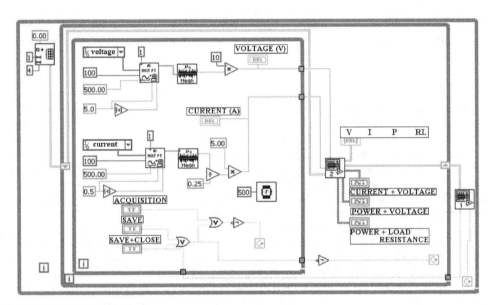

Fig. 6. LabVIEW™ block diagram

3. General principles of data acquisition systems

Data acquisition systems are products and/or processes used to collect informations that can be processed or stored by a computer to document or analyze some phenomenon (Judd, 2008). Measurement systems are used in the data acquisition industry since 1981 and have supplied several million data acquisition channels on an international basis.

Data acquisition systems come in many different PC technology forms to offer flexibility when choosing a measurement system. The organization of information flow in the system is an important problem in designing and operating the measurement. Two aspects are essential for this organization: the kind of transmission in the system (serial, bit-by bit, parallel) and the mode of information exchange between system devices (Nawrocki, 2005).

The functional components of a basic data acquisition are:
- Transducers and sensors;
- Conditioners or circuits standardizing the level of transducer signal to the range of the input voltage of the analog-to digital converter (ADC);
- Analog-to digital converters (ADCs) to convert analog into digital signals;
- Devices for visual display of measurement results (display of digital measurement instruments, oscilloscopes or a computer monitor);
- Computer with dedicated software and memory resources.

More complicated data acquisition systems can be constructed in the hierarchical structure. On the lowest level are subsystems to collect data from physical quantities. The main controller of the system receives processed measuring data and sends commands relating to the execution of a measuring procedure or a set of commands to subsystems.

Analog-to-Digital (A/D) conversion of an analog signal involves two processes:
- Sampling – after this operation signal is represented by a set of values $\{x(kT_s)\}$, drawn at time periods T_s – Figure 7. Sampling frequency should be high enough to provide a number of samples sufficient to reproduce the signal in the analog form. According to the Shannon theorem, the sampling frequency f_s should not be lower than twice the upper frequency f_u of the sampled signal spectrum:

$$f_s > 2f_u \tag{1}$$

Otherwice, the reproduction of the discrete signal recorded yields a distorted analog signal, caused by a too low sampling frequency, phenomenon called aliasing. In order to eliminate aliasing, is utilised a lowpass input filter or an antialiasing filter.
- Quantizing: assigning to every sample a value from a set of N values into which the measurement range is divided. Figure 8 shows a LabVIEW™ application for the quantizing of voltage into an n=3 bit digital signal, the number of quants is $2^3=8$.

The A/D converter is connected to an analog input signal; it measures the analog input and then provides the measurement in digital form suitable for use by a computer. The resolution of an A/D input channel describes the number or range of different possible measurements the system is capable of providing. This specification is almost universally provided in term of "bits". For example, 8-bit resolution corresponds to a resolution of one part in $2^8 - 1$ or 255, 12-bit corresponds to one part in $2^{16} - 1$ or 65,536. To determine the resolution in engineering units, simply divide the range of the input by the resolution. A16-bit input with a 0-10 Volt input range provides 10 V / 2^{16} or 152.6 microvolts. Table 1 provides a comparison of the resolutions for the most commonly used converters in data acquisition systems.

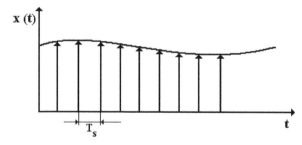

Fig. 7. Signal sampling

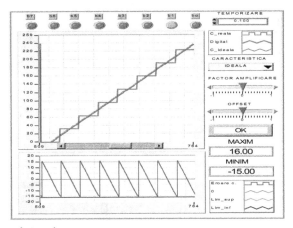

Fig. 8. Quantization of signal

	8-bit	12-bit	16-bit	18-bit	24-bit
Distinct Levels	256	4.096	65.536	262.144	16.777.216
Resolution, ±10 V scale	78.4 mV	4.88 mV	305 μV	76.4 μV	1.192 μV

Table 1. Common Analog-Digital-Converter Resolution

From the point of view of the method of conversion there is a variety of different types of A/D converters used in data acquisition. The most commonly used A/D converters in today's data acquisition products are divided into:
-   Dual Slope/Integrating - with a good attenuation of interferences during the integration process
-   Successive approximation - provides resolutions in the 10 to18-bit range, and depending on the resolution, offers sample rates up to tens of Megasamples per second.
-   Flash – characterized by the shortest conversion time
-   Sigma Delta provides very high resolution, especially in converting continuous signals.

Most multi-channel data acquisition systems are based upon a single A/D converter. A multiplexer is then used between the input channels and the A/D converter. The multiplexer connects a particular input to the A/D, allowing it to sample that channel. Figure 9 depicts a typical, multiplexed input configuration.

The primary disadvantage of this system is that even if the switching and sampling are very fast, the samples are actually taken at different times. The ability to sample inputs at the same instant in time is typically referred to as simultaneous sampling.

There are two ways to achieve simultaneous sampling. The first is to place a separate A/D converter on each channel. They may all be triggered by the same signal and will thus sample the channels simultaneously. The second is to place a device called a sample & hold (S/H) on each input. When commanded to "hold", the S/H effectively freezes its output at that instant and maintains that output voltage until released back into sample mode. Once the inputs have been placed into hold mode, the multiplexed A/D system samples the desired channels. The signal to be sampled will all have been "held" at the same time and so the A/D readings will be of simultaneous samples. The simultaneous sampling configurations are shown in Figure 10.

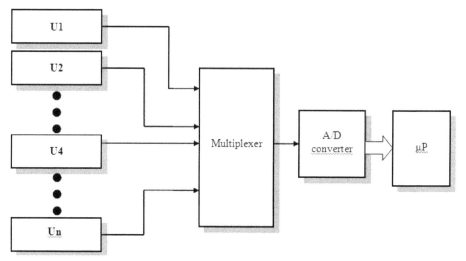

Fig. 9. Typical Multiplexer/ADC DAQ System

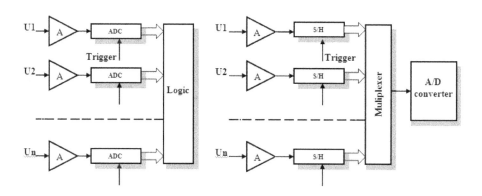

a. using individual ADCs on each input b. using Sample-Hold, MUX, and a single ADC

Fig. 10. Simultaneous Sampling Techniques

A criterion applied to classify data acquisition systems is the transmission type of digital announcements (data, addresses, commands): serial transmission or parallel transmission.

The interface system assures equipment and adjustments of devices attached to the bus. The original PC-bus, ISA-bus, is a simple, robust, and inexpensive interface that has certainly stood the test of time. However, most of today's plug-in board business is based on the PCI bus (or variants such as cPCI and PXI). The external box vendors now have Ethernet and USB standards to work with as well as some less used, but very viable, interfaces such as Firewire, CAN, and perhaps the oldest standard in computing, RS-232.

Software has progressed from the original DOS-based version to hyghly complex applications such as MATLAB and LabVIEW™ that are both easy to use and able to take advantage of today's powerful computers.

Currently, most data acquisition companies provide hardware either at board level or external box and software. PC-based DAQ systems are available with a wide variety of interfaces. Ethernet, PCI, USB, PXI, PCI Express, Firewire, Compact Flash and even the venerable GPIB, RS-232/485, and ISA bus are all popular (Manea & Cepisca, 2007). RS-232 is by far the simplest and least expensive interface. Every PC shipped today can communicate over an RS-232 line. Only a compatible cable and a terminal program are required. RS-232 is limited to about 50 feet, but it can operate over huge distances with a little help from a modem. Only one device can be connected to a single RS-232 port. This interface is typically not electrically isolated from the host computer. For example, if a power line were inadvertently dropped across the RS-232 line, the host computer most likely would be destroyed. As a result, RS-232 may be appropriate for situations in which very few remote I/O systems are needed and where the likelihood of electrical transients is low.

RS-485 allows multiple devices (up to 32) to communicate at half-duplex on a single pair of wires, plus a ground wire (more on that later), at distances up to 1200 meters (4000 feet). Both the length of the network and the number of nodes can easily be extended using a variety of repeater products on the market. Many RS-422/485 protocols were developed in the 1980s for such applications. Profibus, Interbus and CAN are a few. These were designed with proprietary interests in mind, and each subsequently received backing from several vendors in an attempt to develop an open international standard with widespread support.

Ethernet has been around for more than 20 years and has become a commodity in modern business environments. Ethernet also has many characteristics that make it suitable for industrial networked and remote-sensor I/O applications. It combines the low cost of RS-232 with the multidrop capability of RS-422/485 and provides a clear standard for communications. Ethernet has become the medium of choice to communicate management data throughout the enterprise. Interoperability of Ethernet based data acquisition devices from multiple vendors has not always been stellar. However, most Ethernet based data acquisition systems are single vendor and this has not been a major issue in the acquisition space. The LXI Consortium has developed a specification that ensures simple and seamless multi-vendor interoperability.

Gigabit Ethernet is a version of Ethernet that supports 1 Gigabit per second data transfer rates. Boasting the same speed capability as standard Ethernet, the fiber optic implementation extends the range of the system to 2 kilometers. Fiber also provides virtually absolute electrical isolation and has immunity to electrical and magnetic interference.

Firewire is a high speed serial interface. However, at approximately the same time Firewire was being promoted, USB interface was coming on line. At this time, there are a wide variety of data acquisition vendors and products actively promoting USB devices, Firewire success has been confined to the original target market of audio and video.

USB's simple plug-and-play installation, combined with its 480 Mbps data transfer rate, makes it an ideal interface for many data acquisition applications. Also, the popularity of USB in the consumer market has made USB components very inexpensive. USB's 5-meter range is perhaps its largest detraction as it limits the ability to implement remote and distributed I/O systems based on USB.

Today, board level solutions offering 24-bit resolution are now available as, e.g. 6.5 digit DMM boards. On the box side, USB 2.0 is capable of delivering 30 million 16-bit conversions per second and Gigabit Ethernet will handle more than twice that. The internal plug-in slot data transfer rates have increased 10 fold in recent years.

The market of data acquisition equipment is very large, different companies propose new solutions. National Instruments offers several hardware platforms for data acquisition. The most readily available platform is the desktop computer, with PCI DAQ boards that plug into any desktop computer. For distributed measurements, the Compact FieldPoint platform delivers modular I/O, embedded operation, and Ethernet communication. For portable or handheld measurements, National Instruments DAQ devices for USB and PCMCIA work with laptops or Windows Mobile PDAs. In addition, National Instruments has launched DAQ devices for PCI Express, the next-generation PC I/O bus, and for PXI Express, the high-performance PXI bus.

4. Simulations and experimental acquisitions

To determine the operation characteristics of the photovoltaic panels and the panel arrays, the built data acquisition system allows to measure the values of current and voltage, to simultaneously trace characteristics (current-voltage, power-voltage, power-charge resistance) - see figures 11 and 12, to present the measured parameters (during the data acquisition) in tables, to save data into files for future processing.

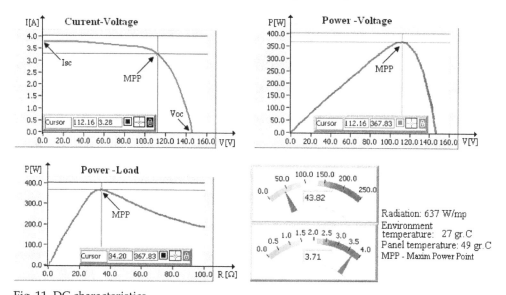

Fig. 11. DC characteristics

For alternative current systems it is necessary to obtain signals with volt-range amplitude to be applied to the input of the data acquisition board. For phase/line voltages there can be used voltage dividers (which do not ensure galvanic isolation) or voltage measurement transformers (ensure galvanic separation). Shunts (current-voltage converters) or current measurement transformers can be used for currents. The use of both voltage dividers and shunts must be done by taking into account the current through the voltage divider, the voltage drop on the shunt, the power dissipation, parasite resistances, self-heating effects and dynamic effects. The use of voltage-current measurement transformers guarantees galvanic isolation of the measuring system but it introduces ratio and angle errors and inadequate perturbation transfer.

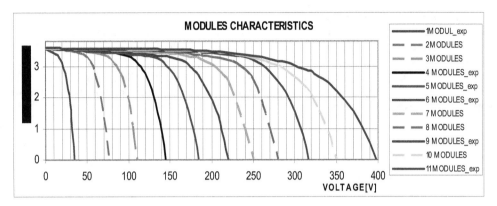

Fig. 12. Current –voltage characteristics. Experimental acquisition (1,4,5,6,9,11 modules) and simulation (2,3,7,8,10 modules)

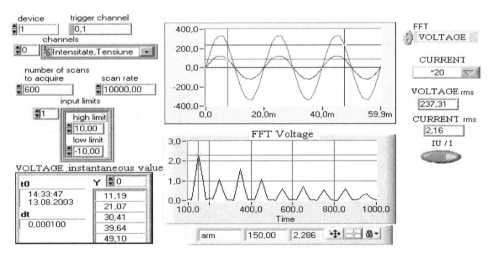

Fig. 13. AC Acquisition Instrument (Phase B)

The solution adopted was to use current and voltage transducers based on the Hall Effect. To obtain a good magnetic sensor the magnetic field is concentrated around the transducer, by using a circular core (used as a flux concentrator and made of a material with high magnetic permeability), which ensures both an increase of the magnetic field in the area of the sensor and the independence from the position of the conductor inside the core. Non-linear behavior of the flux concentrator can be obtained by using an operational amplifier, which injects a compensation current through the reaction loop. Figure 13 presents the front panel of the LabVIEW™ application which acquires and processes data in the AC circuit from the output of the phase B connected inverters.

Figure 14 shows a LabVIEW™ application (Ertugrul, 2002) to determine the characteristics of PV panels, with the possibility of remote monitoring.

4.1 Experimental results in PV system

The following results are obtained with data acquisition system Sunny Data Control. Totally, around 90 channels are recorded permanently. All channels are stored in SBC+ memory for maximum 8 days; after exceeding this interval the data will be lost if not transferred to the computer. Depending on the connected sensors the Sunny Boy Control+ can monitor all inverters and PV panels to ensure an even higher performance and more sophisticated system diagnosis.

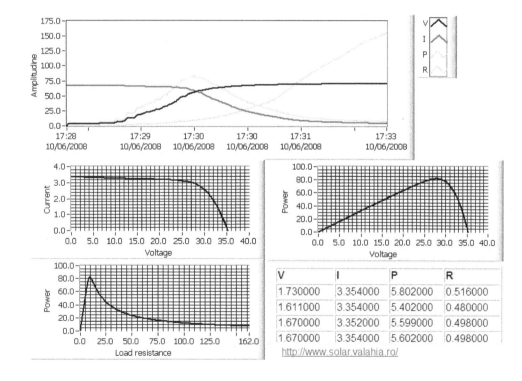

Fig. 14. Remote monitoring of PV system

	1	2	3	3 ~	N		
							Table
U [V]	232.08	229.65	227.57	229.77	0.18	Avg	
I [A]	4.319	2.053	3.073	3.282	0.456		Chart
P [kW]	0.997	0.464	0.695	2.156	0.000	f [Hz]	
S [kVA]	1.002	0.471	0.699	2.173	0.000	49.99	
Q [kVAr]	-0.104	-0.081	-0.076	-0.262	0.000	α u [%]	
P1 [kW]	0.997	0.464	0.696	2.158	0.000	19377.69	
Q1 [kVAr]	-0.055	-0.007	-0.053	-0.115	0.000		
cos φ	1.00	1.00	1.00	1.00	0.03		
PF	0.99	0.98	0.99	0.99	0.01		Init Energy
AP [kWh]	0.134	0.062	0.080	0.276	0.000	Energy	Energy
AS [kVAh]	0.135	0.063	0.081	0.278	0.000		
AQ [kVArh]	-0.015	0.001	-0.009	-0.023	0.000		Max
AP1 [kWh]	0.134	0.062	0.080	0.276	0.000		
AQ1 [kVArh]	-0.008	-0.000	-0.006	-0.013	0.000		Min
APin [kWh]	0.134	0.062	0.080	0.276	0.000		
APout [kWh]	0.000	0.000	0.000	0.000	0.000		Print
AQL [kVArh]	0.000	0.004	0.001	0.000	0.000		
AQC [kVArh]	-0.015	-0.003	-0.010	-0.023	0.000	Break	Store

Fig. 15. Electrical quantities monitored

Figure 15 shows the LabVIEW™ application for the visualization of quantities monitored and the results are presented in Figure 16 (Andrei et al. 2010).

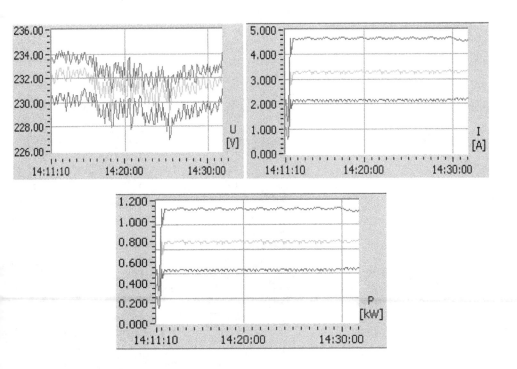

Fig. 16. Results of acquisition

The application allows visualising the phase diagram – Figure 17 - and obtains information on the quality of electricity produced by PV system – Figure 18.

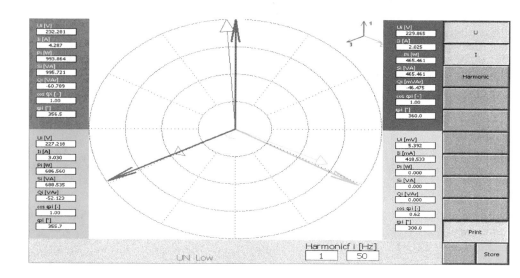

Fig. 17. Phase diagram

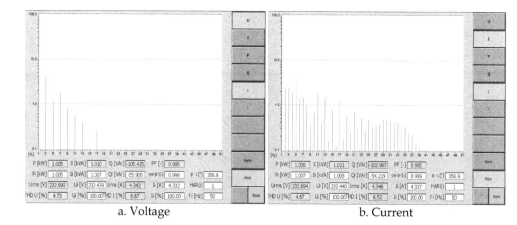

a. Voltage   b. Current

Fig. 18. Harmonics (Phase A)

After processing the data obtained with Sunny Data Control software following characteristics were obtained.

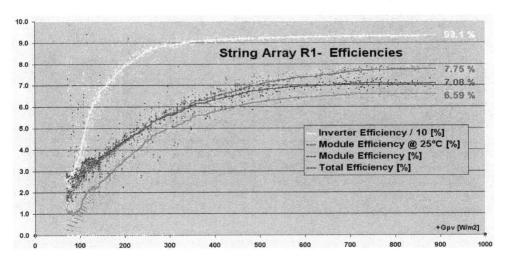

Fig. 19. Efficiency of a panel no.1 according to solar radiation

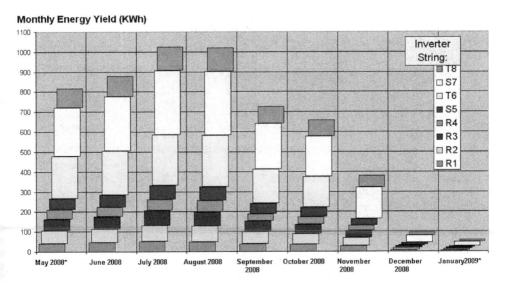

Fig. 20. Monthly electrical energy produced by each rows of modules

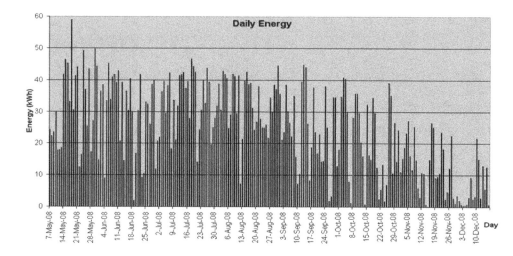

Fig. 21. Total electricity produced by PV system

The data acquisition system allows observation of the link between solar radiation, temperature and solar conversion result – Figures 22 , 23 and 24 .

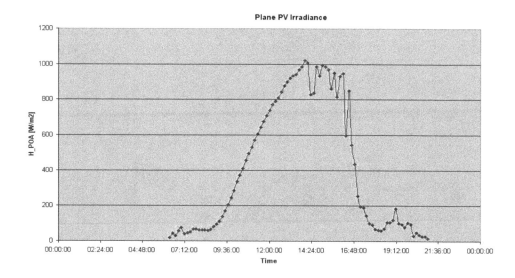

Fig. 22. Solar radiation

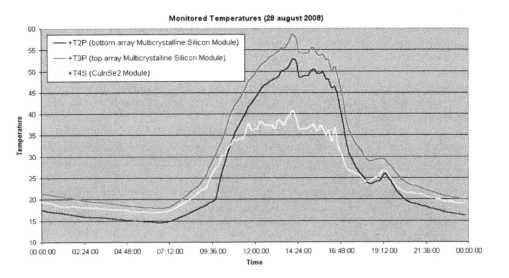

Fig. 23. Temperature evolution of the 3 types of PV modules

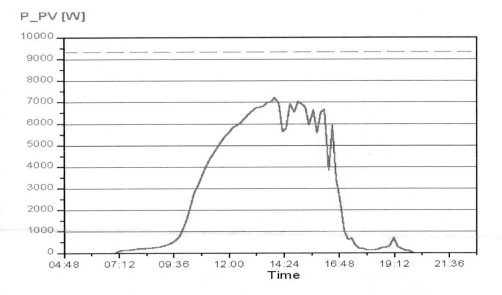

Fig. 24. Variation of total power

5. Conclusion

Even though the costs of installations producing electric energy with PV panels are high compared to the costs of conventional installations, the number of such systems is continuously increasing. It is very important to determine the output characteristics of the PV panels in order to achieve an accurate connection and operation of the device and reduce energy losses.

Monitoring activities follow the operation analysis by periodical reports, papers, synthesis, with the precise aim to make the most accurate decisions to produce electric energy using unconventional sources.

To quantify the potential for performance improvement of a PV system, data acquisition systems has been installed. The importance of this chapter consists in the presentation of a dedicated DAQ used in PV system analysis and real data measurements. The operation is performed by simulations using LabVIEW™.

The information obtained by monitoring parameters, such as voltage, current, power and energies are fed to the PC via the DAQ for analysis. The control interface has been developed by utilizing LabVIEW™ software. The system has been in operation during the last five years and all its units have functioned well.

6. References

Andrei, H.; Dogaru, V.; Chicco, G.; Cepisca, C. & Spertino, F. (2007). Photovoltaic Applications, *Journal of Materials Processing Technology*, 181 (1-3), 2007, 267-273

Andrei, H.; Cepisca, C.; Grigorescu, SD.; Ivanovici, T. & Andrei, P. (2010). Modeling of the PV panels circuit parameters using the 4 - terminals equations and Brune's conditions, *Scientific Bulletin of the Electrical Engineering Faculty*, 10 (1), 2010, 63-67

Ambros, T., et.al. (2004). *Renewable energy*, TEHNICA-INFO, Kishinev

Awerbuch, S. (2002). Energy Diversity and Security in the EU: Mean -Variance Portfolio Analysis of Electricity Generating Mixes, and the Implications for Renewable Sources, *Proceedings of EURELECTRIC Twin Conf. on DG*, pp. 120-125, Brussels, Belgium, 2002

Cepisca, C.; Andrei, H.; Dogaru Ulieru,V. & Ivanovici, T. (2004). Simulation and data acquisition of the photovoltaic systems using LabVIEW™, *Proceedings of ICL 2004*, pp. 80-84, Villach, Austria, 2004

Dogaru Ulieru,V.; Cepisca, C. & Ivanovici, T. (2009). Data Acquisition in Photovoltaic Systems, *Proceedings of 13th WSEAS International Conference on Circuits, Systems, Communications and Computers*, pp. 234-238, Rodos Island, Greece, July 22-24, 2009

Ertugrul, N. (2002). *LabVIEW™ for electric circuits, machines, drives and laboratories*, Ed. Prentice Hall, New York

Judd,B. (2008). Everything You Ever Wanted to Know about Data Acquisition, In: *United Electronic Industries*, 2008, available from *www.ueidaq.com*

Manea,F. & Cepisca, C. (2007). PHP+Apache+Testpoint -An original way for having remote control over any type of automation, *Scientific Bulletin UPB, Series C Electrical Engineering*, 69 (2), 2007, 85-92

Nawrocki, W. (2005). *Measurement System and Sensors*, Artech House, London

Szekely, I. (1997). *Systems for data acquisition and processing*, Ed. Mediamira, Cluj–Napoca

Vasile, N. (2009). Players on the market in renewable energy, *Round Table - renewable sources of energy between the European Directive 77/2001 and reality*, Bucharest, Romania, May 2009

11

Optimum Design of a Hybrid Renewable Energy System

Fatemeh Jahanbani and Gholam H. Riahy
Electrical Engineering Department, Amirkabir University of Technology
Iran

1. Introduction

In Iran, 100% of the region populated with more than 20 families is electrified. For the other regions the electrification will be done. These regions almost are rural and remote areas. For utility company it is important that electrification be done with the least cost.

Many alternative solutions could be used for this goal (decreasing the cost). Using renewable energy system is one of the possible solutions. A growing interest in renewable energy resources has been observed for several years, due to their pollution free energy, availability, and continuity. In practice, use of hybrid energy systems can be a viable way to achieve trade-off solutions in terms of costs. Photovoltaic (PV) and wind generation (WG) units are the most promising technologies for supplying load in remote and rural regions [Wang et al., 2007]. Therefore, in order to satisfy the load demand, hybrid energy systems are implemented to combine solar and wind energy units and to mitigate or even cancel out the power fluctuations. Energy storage technologies, such as storage batteries (SBs) can be employed. The proper size of storage system is site specific and depends on the amount of renewable energy generation and the load.

Many papers are discussed on design of hybrid systems with the different components. Also, various optimization techniques are used by researchers to design hybrid energy system in the most cost effective way.

Rahman and Chedid give the concept of the optimal design of a hybrid wind–solar power system with battery storage and diesel sets. They developed linear programming model to minimize the average production cost of electricity while meeting the load requirements in a reliable manner, and takes environmental factors into consideration both in the design and operation phases [Chedid et al., 1997]. In [Kellogg et al, 1996], authors proposed an iterative technique to find the optimal unit sizing of a stand-alone and connected system. In 2006 is presented a methodology for optimal sizing of stand-alone PV/WG systems using genetic algorithms. They applied design approach of a power generation system, which supplies a residential household [Koutroulis et al, 2006]. In [Ekren, 2008], authors used the response surface methodology (RSM) in size optimization of an autonomous PV/wind integrated hybrid energy system with battery storage. In [Shahirinia, 2005], an optimized design of stand-alone multi sources power system includes sources like, wind farm, photovoltaic array, diesel generator, and battery bank based on a genetic algorithm is presented. Also, authors in [Koutroulis et al, 2006, Tina, 2006] used multi-objective genetic algorithm, in order to calculate reliability/cost implications of hybrid PV/wind energy system in small isolated power systems. Yang developed a novel optimization sizing model for hybrid solar–wind power generation system [Yang et al., 2007]. In [Terra, 2006] an automatic multi-objective optimization procedure base on fuzzy logic for grid connected HSWPS design is described. In some later works, PSO is successfully implemented for optimal sizing of hybrid stand-alone power systems, assuming continuous and reliable supply of the load [Lopez, 2008, Belfkira, 2008]. Karki and Billinton presented a Monte-Carlo simulation approach to calculate the reliability index [Karki et al., 2001] and Kashefi presented a method for assessment of reliability basis on binominal distribution function for hybrid PV/wind/fuel cell energy system that is used in this study [Wang et al., 2007].

As previous studies shown, renewable energies are going to be a main substitute for fossil fuels in the coming years for their clean and renewable nature [Sarhaddi et al., 2010]. Photovoltaic solar and wind energy conversion systems have been widely used for electricity supply in isolated locations that are far from the distribution network.

The future of power grids is expected to involve an increasing level of intelligence and integration of new information and communication technologies in every aspect of the electricity system, from demand-side devices to wide-scale distributed generation to a variety of energy markets.

In the smart grid, energy from diverse sources is combined to serve customer needs while minimizing the impact on the environment and maximizing sustainability. In addition to nuclear, coal, hydroelectric, oil, and gas-based generation, energy will come from solar, wind, biomass, tidal, and other renewable sources. The smart grid will support not only centralized, large-scale power plants and energy farms but residential-scale dispersed distributed energy sources [Santacana et al., 2010].

Being able to accommodate distributed generation is an important characteristic of the smart grid. Because of mandated renewable portfolio standards, net metering requirements and a desire by some consumers to be green, there is an increasing need to be prepared to be able to interconnect generation to distribution systems, especially renewable generation such as photovoltaic, small wind and land fill gas powered generation [Saint, 2009].

The future electric grid will invariably feature rapid integration of alternative forms for energy generation. As a national priority, renewable energy resources applications to offset the dependence on fossil fuels provide green power options for atmospheric emissions curtailment and provision of peak load shaving are being put in policy [Santacana et al., 2010].

Fortunately, Iran is a country with the adequate average of solar radiation and wind speed for setting up a hybrid power generation e.g. the average of wind speed and perpendicular solar radiation were recorded for Ardebil province is 5.5945 m/s and 203.1629 W/m^2 respectively in a year.

In this study, an optimal hybrid energy generation system including of wind, photovoltaic and battery is designed. The aim of design is to minimize the cost of the stand-alone system over its 20 years of operation. The optimization problem is subject to economic and technical constraints. Figure1 show the framework of activities in this study.

The generated power by wind turbine and PV arrays are depended on many parameters that the most effectual of them are wind speed, the height of WTs hub (that affects the wind speed), solar radiations and orientation of PV panels. In certain region, the optimization variables are considered as the number of WTs, number of PV arrays, installation angle of PV arrays, number of storage batteries, height of the hub and sizes of DC/AC converter. The goal of this study is optimal design of hybrid system for the North West of Iran (Ardebil province). The data of hourly wind speed, hourly vertical and horizontal solar radiation and load during a year are measured in the region. This region is located in north-west of Iran and there are some villages far from the national grid. The optimization is carried out by Particle Swarm Optimization (PSO) algorithm. The objective function is cost with considered economical and technical constraints. Three different scenarios are considered and finally economical system is selected.

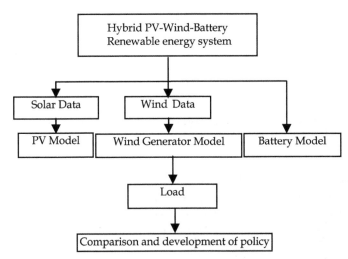

Fig. 1. The framework of activities

This study is organized as follows: section 2 describes the modeling of system components. The reliability assessment is discussed in section 3. Problem formulation and operation strategy are explained in section 4 and 5, respectively. In the next section, is dedicated to particle swarm optimization. Simulation and results are summarized in section 7. Finally, section 8 is devoted to conclusion.

2. Description of the hybrid system

The increasing energy demand and environmental concerns aroused considerable interest in hybrid renewable energy systems and its subsequent development.

The generation of both wind power and solar power is very dependent on the weather conditions. Thus, no single source of energy is capable of supplying cost-effective and reliable power. The combined use of multiple power resources can be a viable way to achieve trade-off solutions. With combine of the renewable systems, it is possible that power fluctuations will be incurred. To mitigate or even cancel out the fluctuations, energy storage technologies, such as storage batteries (SBs) can be employed [Wang et al., 2009].

The proper size of storage system is site specific and depends on the amount of renewable generation and the load. The needed storage capacity can be reduced to a minimum when a proper combination of wind and solar generation is used for a given site [Kellogg, 1996]. The hybrid system is shown in Fig. 2. In the following sections, the model of components is discussed.

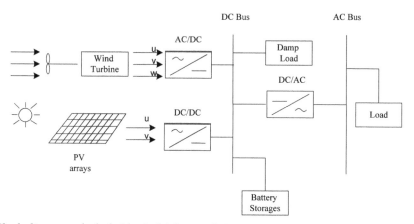

Fig. 2. Block diagram of a hybrid wind/photovoltaic generation unit

2.1 The wind turbine

Choosing a suitable model is very important for wind turbine power output simulations. The most simplified model to simulate the power output of a wind turbine could be calculated from its power-speed curve. This curve is given by manufacturer and usually describes the real power transferred from WG to DC bus.

The model of WG is considered BWC Excel-R/48 (see Fig. 3) [Hakimi et al., 2009]. It has a rated capacity of 7.5 kW and provides 48 V dc as output. The power of wind turbine is described in terms of the wind speed according to Eq. 1.

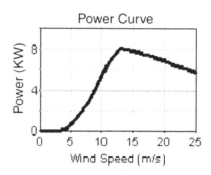

Fig. 3. Power output characteristic of BWC Excel R/48 versus wind speed [Hakimi, 2009].

$$P_W = \begin{cases} 0 & v_W \leq v_{ci}, v_W \geq v_{co} \\ P_{W_{max}} \times \left(\dfrac{v_W - v_{ci}}{v_r - v_{ci}} \right)^m & v_{ci} \leq v_W \leq v_r \\ P_{W_{max}} + \dfrac{P_f - P_{W_{max}}}{v_{co} - v_r} \times (v_W - v_r) & v_r \leq v_W \leq v_f \end{cases} \qquad (1)$$

where $P_{WG_{max}}$, P_f are WG output power at rated and cut-out speeds, respectively. Also, v_r, v_{ci}, v_{co} are rated, cut-in and cut-out wind speeds, respectively. In this study, the exponent m is considered 3. In the above equation, v_W refers to wind speed at the height of WG's hub. Since, v_W almost is measured at any height (here, 40 m), not in height of WGs hub, is used Eq. (2) to convert wind speed to installation height through power law [Borowy et al., 1996]:

$$v_W = v_W^{measure} \times \left(\frac{h_{hub}}{h_{measure}} \right)^\alpha \tag{2}$$

where α is the exponent law coefficient. α varies with such parameters as elevation, time of day, season, nature of terrain, wind speed, temperature, and various thermal and mechanical mixing parameters. The determination of α becomes very important. The value of 0.14 is usually taken when there is no specific site data (as here) [Yang et al., 2007].

2.2 The photovoltaic arrays (PVs)

Solar energy is one of the most significant renewable energy sources that world needs. The major applications of solar energy can be classified into two categories: solar thermal system, which converts solar energy to thermal energy, and photovoltaic (PV) system, which converts solar energy to electrical energy. In the following, the modeling of PV arrays is described.

For calculating the output electric power of PVs, perpendicular radiation is needed. When the hourly horizontal and vertical solar radiation is available (as this study), perpendicular radiation can be calculated by Eq. (3):

$$G(t, \theta_{PV}) = G_V(t) \times \cos(\theta_{PV}) + G_H(t) \times \sin(\theta_{PV}) \tag{3}$$

where, $G_V(t)$ and $G_H(t)$ are the rate of vertical and horizontal radiations in the tth step-time (W/m²), respectively. The radiated solar power on the surface of each PV array can be calculated by Eq. (4):

$$P_{pv} = \frac{G}{1000} \times P_{pv,rated} \times \eta_{MPPT} \tag{4}$$

where, G is perpendicular radiation at the arrays' surface (W/m²). $P_{pv,rated}$ is rated power of each PV array at $G = 1000(W / m^2)$ and η_{MPPT} is the efficiency of PV's DC/DC converter and Maximum Power Point Tracking (MPPT).

2.3 The storage batteries

Since both wind and PVs are intermediate sources of power, it is highly desirable to incorporate energy storage into such hybrid power systems. Energy storage can smooth out the fluctuation of wind and solar power and improve the load availability [Borowy et al., 1996].

When the power generated by WGs and PVs are greater than the load demand, the surplus power will be stored in the storage batteries for future use. On the contrary, when there is any deficiency in the power generation of renewable sources, the stored power will be used to supply the load. This will enhance the system reliability.

In the state of charge, amount of energy that will be stored in batteries at time step of t is calculated:

$$E_B(t) = E_B(t-1) + \left(\left(P_w + P_{pv} \right)(t) - P_{Load}(t) / \eta_{inv.} \right) \eta_{Bat} \tag{5}$$

In addition, Eq. 6 will calculate the state of battery discharge at time step of t:

$$E_B(t) = E_B(t-1) + \left(P_{Load}(t) / \eta_{inv.} - \left(P_w + P_{pv} \right)(t) \right) \eta_{Bat} \tag{6}$$

where, $E_B(t)$, $E_B(t-1)$ are the stored energy of battery in time step of t and $(t-1)$. P_w, P_{pv} are the generated power by wind turbines and PV arrays, $P_{Load}(t)$ is the load demand at time step of t and η_{Bat} is the efficiency of storage batteries.

2.4 The power inverter

The power electronic circuit (inverter) used to convert DC into AC form at the desired frequency of the load. The DC input to the inverter can be from any of the following sources:
1. DC output of the variable speed wind power system
2. DC output of the PV power system
In this study, supposed the inverter's efficiency is constant for whole working range of inverter (here 0.9).

3. The reliability assessment

A widely accepted definition of reliability is as follows [Billinton, 1992]: "Reliability is the probability of a device performing its purpose adequately for the period of time intended under the operating conditions encountered". In the following sections, reliability indices and reliability model that is used in this study is described.

3.1 Reliability indices

Several reliability indices are introduced in literature [Billinton, 1994, XU et al., 2005]. Some of the most common used indices in the reliability evaluation of generating systems are Loss of Load Expected (LOLE), Loss of Energy Expected (LOEE) or Expected Energy not Supplied (EENS), Loss of Power Supply Probability (LPSP) and Equivalent Loss Factor (ELF).
In this study, ELF is chosen as the main reliability index. On the other word, the ELF index is chosen as a constraint that must be satisfied but it could be possible to calculate the other indexes as is done in this study (such as EENS, LOLE and LOEE indexes).
ELF is ratio of effective load outage hours to the total number of hours. It contains information about both the number and magnitude of outages. In the rural areas and stand-alone applications (as this study), ELF<0.01 is acceptable. Electricity supplier aim at 0.0001 in developed countries [Garcia et al., 2006]:

$$ELF = \frac{1}{H} \sum_{h=1}^{H} \frac{E(Q(h))}{D(h)} \tag{7}$$

where, $Q(h)$ and $D(h)$ are the amount of load that is not satisfied and demand power in h^{th} step, respectively and H is the number of time steps (here H=8760).

In this study, the reliability index is calculated from component's failure, that is concluding of wind turbine, PV array, and inverter failure.

3.2 System's reliability model

As mentioned, outages of PV arrays, wind turbine generators, and DC/AC converter are taken into consideration. Forced outage rate (FOR) of PVs and WGs is assumed to be 4% [Karki et al., 2001]. So, these components will be available with a probability of 96%. Probability of encountering each state is calculated by binomial distribution function [Nomura 2005].

For example, given n_{WG} fail out of total N_{WG} installed WGs, and n_{PV} fail out of total N_{PV} installed PV arrays are failed, the probability of encountering this state is calculated as follows:

$$f_{ren}\left(n_{WG}^{fail}, n_{PV}^{fail}\right) = \left[\binom{n_{WG}^{fail}}{N_{WG}} \times A_{WG}^{N_{WG}-n_{WG}^{fail}} \times \left(1-A_{WG}\right)^{n_{WG}^{fail}}\right] \times \left[\binom{n_{PV}^{fail}}{N_{PV}} \times A_{PV}^{N_{PV}-n_{PV}^{fail}} \times \left(1-A_{PV}\right)^{n_{PV}^{fail}}\right]$$ (8)

The outage probability of other components is negligible. But, because, DC/AC converter is the only single cut-set of the system reliability diagram, the outage probability of it is taken consideration (it's FOR is considered 0.0011 [Kashefi et al., 2009]).

In [Kashefi et al., 2009] an approximate method is used that proposed all the possible states for outages of WGs and PV arrays to be modeled with an equivalent state. This idea is modeled by Eq. 7.

$$E[P_{ren}] = N_{WG} \times P_{WG} \times A_{WG} + N_{PV} \times P_{PV} \times A_{PV}$$ (9)

4. Problem formulation

The economical viability of a proposed plant is influence by several factors that contribute to the expected profitability. In the economical analysis, the system costs are involved as:
- Capital cost of each component
- Replacement cost of each component
- Operation and maintenance cost of each component
- Cost costumer's dissatisfaction

It is desirable that the system meets the electrical demand, the costs are minimized and the components have optimal sizes. Optimization variables are number of WGs, number of PV arrays, installation angle of PV arrays, number of storage batteries, and sizes of DC/AC converter. For calculation of system cost, the Net Present Cost (NPC) is chosen.

For optimal design of a hybrid system, total costs are defined as follow:

$$NPC_i = N_i \times (CC_i + RC_i \times K_i + O \& MC_i \times 1 / CRF(ir, R))$$ (10)

where N may be number (unit) or capacity (kW), CC is capital cost (US\$/unit), $O\&MC$ is annual operation and maintenance cost (US\$/unit-yr) of the component. R is Life span of project, ir is the real interest rate (6%). CRF and K are capital recovery factor and single payment present worth, respectively.

$$ir = \frac{(ir_{nominal} - f)}{(1 + f)} \tag{11}$$

$$CRF(ir, R) = \frac{ir(1 + ir)^R}{(1 + ir)^R - 1} \tag{12}$$

$$K_i = \sum_{n=1}^{y_i} \frac{1}{(1 + ir)^{n \times L_i}} \tag{13}$$

4.1 The cost of loss of load

In this study, cost of electricity interruptions is considered. The values found for this parameter are in the range of 5-40 US$/kWh for industrial users and 2-12 US$/kWh for domestic users [Garcia et al., 2006]. In this study, the cost of customer's dissatisfaction, caused by loss of load, is assumed to be 5.6 US$/kWh [Garcia et al., 2006].

Annual cost of loss of load is calculated by:

$$NPC_{loss} = LOEE \times C_{loss} \times PWA \tag{14}$$

where, C_{loss} is cost of costumer's dissatisfaction (in this study, US$5.6/kWh). Now, the objective function with aim to minimize total cost of system is described:

$$Cost = \sum_i NPC_i + NPC_{loss} \tag{15}$$

where i indicates type of the source, wind, PV, or battery. To solve the optimization problem, all the below constraints have to be considered:

$$\begin{aligned}
&0 \le N_i < N_{max} \\
&10 \le H_{hub} \le 20 \\
&0 \le \theta_{PV\&PVT} \le \pi/2 \\
&E_{bat_{min}} \le E_{bat} \le E_{bat_{max}} \\
&E[ELF] \le ELF_{max}
\end{aligned} \tag{16}$$

The last constraint is the reliability constraint. Equivalent Loss Factor is ratio of effective load outage hours to the total number of hours. In the rural areas and stand-alone applications (as this study), ELF<0.01 is acceptable [Tina, 2006]. For solving the optimization problem, particle swarm algorithm has been exploited.

5. Operation strategy

The system is simulated for each hour in period of one year. In each step time, one of the below states can occur:

- If the total power generated by PV arrays and WGs are greater than demanded load, the energy surplus is stored in the batteries until the full energy is stored. The remainder of the available power is consumed in the dump load.

-   If the total power generated by PV arrays and WGs are less than demanded load, shortage power would be provided from batteries. If batteries could not provide total energy that loads demanded, the load will be cut.
-   If the total power generated by PV arrays and WGs are equal to the demanded load, the storage capacity remains unchanged and all of the generated power will be consumed at the load.

By consideration these states and all the constraints, the optimal hybrid system is calculated.

6. Optimization method

For size optimization of components PSO algorithm is used. Direct search method (traditional optimization method) heavily depends on good starting points, and may fall into local optima. On the other hand, as a global method for solving both constrained and unconstrained optimization problems based on natural evolution, the PSO can be applied to solve a variety of optimization problems that are not well suited for standard optimization algorithms. Moreover, the GA can also be employed to solve a variety of optimization problems. Compared to GA, the advantages of PSO are that PSO is easy to implement and there are few parameters to adjust. PSO has been successfully applied in many areas.

6.1 The PSO algorithm

Particle swarm optimization was introduced in 1995 by Kennedy and Eberhart. The following is a brief introduction to the operation of the PSO algorithm. The particle swarm optimization (PSO) algorithm is a member of the wide category of swarm intelligence methods for solving global optimization problems. PSO is an evolutionary algorithm technique through individual improvement plus population cooperation and competition which is based on the simulation of simplified social models, such as bird flocking, fish schooling and the swarm theory [Jahanbani et al., 2008].

Each individual in PSO, referred as a particle, represents a potential solution. In analogy with evolutionary computation paradigms, a swarm is similar to population, while a particle is similar to an individual.

In simple terms, each particle is flown through a multidimensional search space, where the position of each particle is adjusted according to its own experience and that of its neighbors.

Assume x and v denote a particle position and its speed in the search space. Therefore, the ith particle can be represented as $x_i = [x_{i_1}, x_{i_2}, ..., x_{i_d}, ..., x_{i_N}]$ in the N-dimensional space. Each particle continuously records the best solution it has achieved thus far during its flight. This fitness value of the solution is called *pbest*. The best previous position of the ith particle is memorized and represented as:

$$pbest_i = [pbest_{i_1}, pbest_{i_2}, ..., pbest_{i_d}, ..., pbest_{i_N}] \tag{17}$$

The global best *gbest* is also tracked by the optimizer, which is the best value achieved so far by any particle in the swarm. The best particle of all the particles in the swarm is denoted by $gbest_d$. The velocity for particle i is represented as $v_i = [v_{i_1}, v_{i_2}, ..., v_{i_d}, ..., v_{i_N}]$.

The velocity and position of each particle can be continuously adjusted based on the current velocity and the distance from $pbest_{id}$ to $gbest_d$:

$$v_i(t+1) = \omega(t)v_i(t) + c_1r_1(P_i(t) - X_i(t)) + c_2r_2(G(t) - X(t)) \tag{18}$$

$$X_i(t+1) = X_i(t) + \chi v_i(t+1) \tag{19}$$

where c_1 and c_2 are acceleration constants and r_1 and r_2 are random real numbers drawn from [0,1]. Thus the particle flies trough potential solutions toward $P_i(t)$ and $G(t)$ in a navigated way while still exploring new areas by the stochastic mechanism to escape from local optima.

Since there was no actual mechanism for controlling the velocity of a particle, it was necessary to impose a maximum value V_{max}, which controls the maximum travel distance in each iteration to avoid this particle flying past good solutions. Also after updating the positions, it must be checked that no particle violates the boundaries of search space. If a particle has violated the boundaries, it will be set at boundary of search space [Jahanbani et al., 2008].

In Eq. (20), $\omega(t)$ is employed to control the impact of the previous history of velocities on the current one and is extremely important to ensure convergent behavior. It is exposed completely in the following section. $\omega(t)$ is the constriction coefficient, which is used to restrain velocity. χ is constriction factor which is used to limit velocity, here $\chi = 0.7$.

7. Simulation results

The first goal of each planning in electrical network is that the system meets the demand. For satisfying this goal, the cost of costumer's dissatisfaction is considered as well as the other costs. Flowchart of the proposed optimization methodology is shown in Fig. 4. The hourly data of wind speed, vertical and horizontal solar radiation and residential load during a year is plotted in Fig. 5, Fig. 6 and Fig. 7, respectively. The data that used in this study is the data of Ardebil convince that is located in the North West of Iran (latitude: 38°17′, longitude: 48°15′, altitude: 1345 m). The peak load is considered as 50 kW. In table 1, data that used in the simulation are listed.

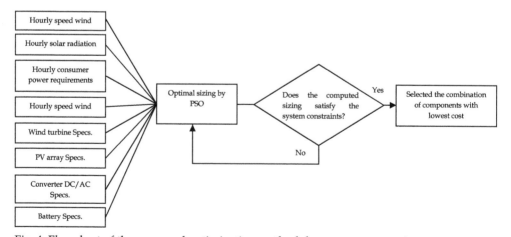

Fig. 4. Flowchart of the proposed optimization methodology.

System parameters	values	System parameters	values
Efficiency of SB	85%	Replacement price of PV array	6000 US$/unit
Efficiency of inverter	90%	Replacement price of SB	700 US$/unit
Life span of project	20	Replacement price of inverter	750 US$/unit
Life span of WTG and PV	20	OM costs of inverter	8 US$/unit-yr
Life span of SB	10	Cut-in wind speed	3 m/s
Life span of inverter	15	Rated wind speed	13 m/s
PV array price	7000 US$/unit	Cut-out wind speed	25 m/s
WTG price	19400 US$/unit	Rated WTG power	7.5 kW
Inverter price	800 US$/unit	Minimum storage level of battery	3 kWh
Replacement price of WTG	15000 US$/unit	Maximum total SB capacity	40 kWh

Table 1. Data used for simulation program [Tina et al., 2006, Khan et al., 2005]

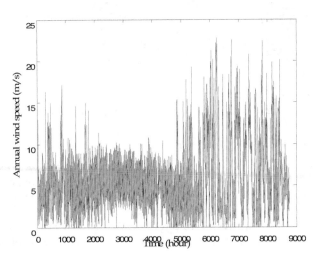

Fig. 5. Hourly wind speed during a year.

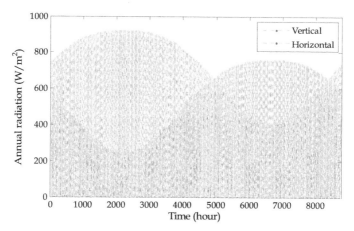

Fig. 6. Hourly vertical and horizontal solar radiation during a year.

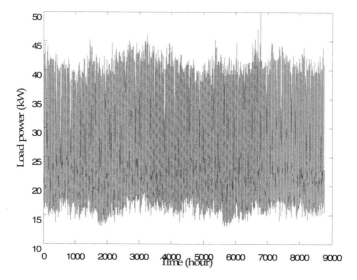

Fig. 7. Hourly residential load during a year.

It is noticeable that the technical constraint, related to system reliability, is expressed by the equivalent loss factor. The reliability index is calculated from component's failure, that includes wind turbine, PV array, battery and inverter failure. The power generated by each wind turbine and PV array can be derived by Eq. (1) and Eq. (3), respectively. The total power that can be generated with N_{WG} wind turbines and N_{PV} PV arrays that n_{WG} and n_{PV} of all wind turbines and PV arrays are out of work, respectively, will be calculated as follows:

$$P_{ren} = (N_{WG} - n_{WG}) \times A_{WG} \times P_{WG} + (N_{PV} - n_{PV}) \times A_{PV} \times P_{PV} \tag{20}$$

As previously mentioned, the reliability constraint is considered as the penalty factor in the objective function. To consider the constraint of reliability in Eq. (16), the excess amount of inequality constraint is multiplied by 1010 and then, this additional cost is added to the objective function in Eq. (15). With this method, the NPC of the system that couldn't satisfy the reliability constraint will increase, and then this system would not be chosen as the best economic system.

One of the best methods in the planning area is using scenario method. To choose the best plan (the minimum cost) different scenarios is implemented. In this study, the optimal size of components for hybrid system is calculated in three scenarios based on proposed approach. These systems are PV/battery system, wind/battery system and PV/wind/battery system. For each system the minimum cost and reliability indices is calculated. The results are shown in the following.

As mentioned before, in this study particle swarm optimization algorithm is used for optimal sizing of system's components. Each particle has 6 variables that are defined as below:

Number of wind turbine	Number of PV array	Angle of PV	Battery capacity	Height of hub	Inverter capacity

Fig. 8. A typical vector for a particle

Each population consists of 30 particles that are calculated for 120 iterations. The fitness function is defined in Eq. (15). It must be considered that if the costs of loss of load are more expensive than the cost of bigger system, the bigger system will be chosen because it is economically reasonable.

7.1 Scenario I: Wind/PV/battery hybrid system

In this case, a stand-alone hybrid system is considered that is including of wind and PV energy sources and storage batteries. Convergence curves of the PSO algorithm, for five independent runs, are shown in Fig. 9. It is observed that the algorithm converges almost to the same optimal value.

Hourly generated power of PV arrays, WGs is shown in Fig. 10 that could be comparing with load. The hourly expected amount of stored energy in the battery is shown in Fig. 10, too. It is significant that reliable supply of the load at each time step, strongly, depends on the amount of the stored energy. When stored energy in battery reaches its minimum allowable limit, if renewable system cannot satisfy the load, the load will not be supplied. On the other hand, if renewable system can satisfy the load, the extra generated energy will be saved in the battery (and battery is in the state of charge). It is worth pointing out that when the battery has the maximum charge its energy will not increase anymore.

In Fig. 11, the hourly reliability indices in the year are plotted. The amount of hourly demand and load pattern is another important factor in reliability assessment of the system. The size of each component is also calculated and is shown in table 2.

As shown in the above figures, each time step could be analyzed. For example, at around of 6500th time step, the power that is generated by PV arrays and wind turbines is decreased (Fig. 10) and it is not enough to satisfy the load. Also, the energy that saved in the batteries in this step is around the minimum allowable level. So, some of the demand is lost and ELF index is equal to 0.5 (Fig. 11).

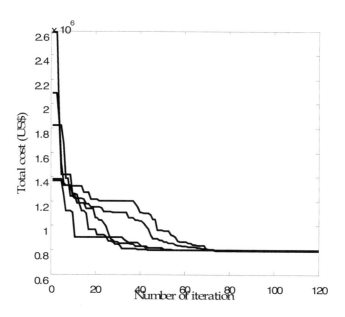

Fig. 9. Convergence of the optimization algorithm

N_{WG}	N_{PV}	N_{Bat}	P_{inv} (kW)	θ_{PV}	$H_{hub}(m)$
1	89	12	44	52.61	15.85

Table 2. Optimal combination for hybrid system

ELF	LOEE (kWh/yr)	EENS	LOLE(h/yr)
0.0036	759.49	0.0034	64.57

Table 3. Reliability indices of PV-wind-battery system

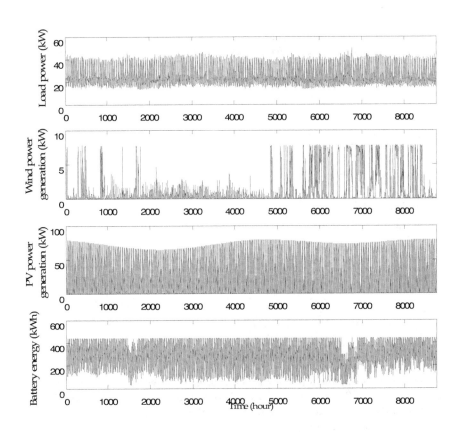

Fig. 10. Hourly generated power of PV arrays, WGs and hourly expected amount of stored energy in the battery during a year.

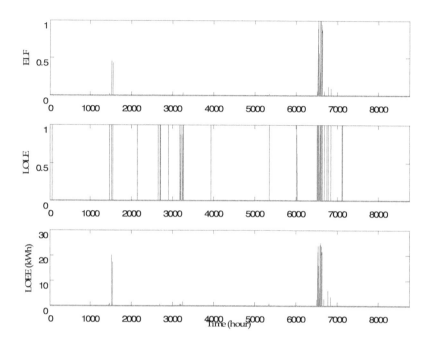

Fig. 11. Hourly reliability indices during a year

In this scenario, the mean of ELF index in the year is 0.002 which is less than the maximum ELF tconstraint (0.01). So, this system would not pay for penalty cost. The NPC, which is calculated for this case, would be equal to 1.29769 MUSD that 31272.02 USD of this cost would be for costumer's dissatisfaction.

7.2 Scenario II: Wind/battery system

This system is including of wind source energy and storage batteries. The optimal size of system components is presented in table 4. In this case, the reliability constraint is activated, so it is fixed on maximum allowable value. Because of this, the NPC of system is increased and is reached up to 2.25009 MUS$. The generated power by wind turbines and amount of energy in storage system is shown in Fig. 12.

As mentioned, in this system, the ELF index is reached to 0.0063 which satisfy the inequality constraint of reliability constraint. Thus, it must not pay the penalty cost and the costumer's dissatisfaction cost would be 0.032424 MUS$.

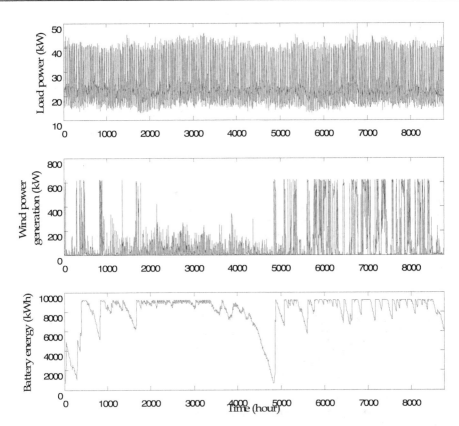

Fig. 12. Hourly generated power of WGs and hourly expected amount of stored energy in the battery during a year.

N_{WG}	N_{PV}	N_{Bat}	P_{inv} (kW)	θ_{PV}	$H_{hub}(m)$
80	-	230	44.5	-	19

Table 4. Optimal combination in wind-Battery system

ELF	LOEE (kWh/yr)	EENS	LOLE(h/yr)
0.0063	1315.6	0.0063	77.56

Table 5. Reliability indices of wind-battery system

7.3 Scenario III: PV/Battery systems

The last scenario is a system including of PV source energy and storage batteries. The size of system components is shown in table 6. Total cost and ELF index corresponding to this case are 0.803237 MUS$ and 0.0022 respectively that, 0.032423 MUS$ would be paid as costumer's dissatisfaction cost. The output power of PV arrays and battery energy is shown in Fig. 13.

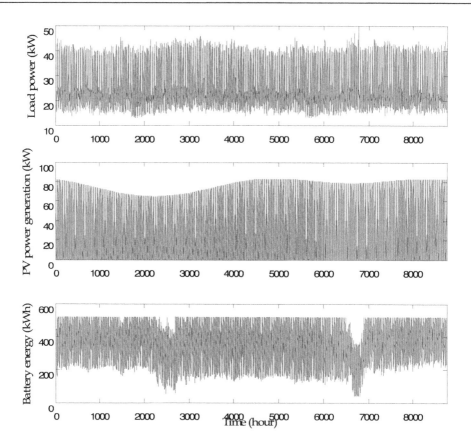

Fig. 13. Hourly generated power of PV arrays and hourly expected amount of stored energy in the battery during a year.

N_{WG}	N_{PV}	N_{Bat}	P_{inv} (kW)	θ_{PV}	$H_{hub}(m)$
-	96	13	44.3	56.74	11

Table 6. Optimal combination in PV-battery system

ELF	LOEE (kWh/yr)	EENS	LOLE(h/yr)
0.0022	504.79	0.0024	49.59

Table 7. Reliability indices of PV-battery system

With compare of these scenarios together, it could be seen, the number of batteries in wind-battery system is more than the hybrid system and PV-battery system. That's reasonable because in this region (and almost all of regions) fluctuations of wind are more than the fluctuations in radiation, so, when the wind turbine is used, we needed to add more storage system to be sure that the load would be met in all steps. Also, in this region, the peak load is nearer to the peak of PV generation compared with the peak of wind generation. On the other hand, Because of this, the reliability index in the system with wind turbine is less than the system with PV array and subsequently, because of increase in costumer's dissatisfaction cost, the cost of the system with wind turbine is more than the system with PV array.

In any of the scenarios, the reliability constraint is not activated and each scenario will be able to satisfy the inequality constraints without penalty cost. As mentioned before, it is possible to analyze each point of the results to investigate the relation between changes of wind, solar radiation and load with ELF index and charging and discharging states of batteries.

8. Conclusions

In this paper, a hybrid generation system is designed for a 20-year period of operation for the North west of Iran. The major components are PV arrays and wind turbines. The major advantage of these components is that when used together, the reliability of the system is enhanced. Additionally, the size of storage systems can be reduced as there is less reliance on one method of power production. Often, when there is no sun, there is plenty of wind and vice versa.

The number of components is directly dependent on the load pattern. The region has a cold climate thus electricity demand in summer is not significantly more than the demand in winter. This is an advantage because the needed battery will decrease and the design system will be more economical.

In this study, the batteries are employed as the energy storage system. Optimal combination of components is achieved by particle swarm optimization. The optimization problem is subject to economical and technical constraints. Best configuration with considered reliability constraint is achieved and the system is simulated.

In the future work of this study, uncertainty factors such as generator failures and renewable power availability will also be taken into account in calculating system reliability indexes.

9. References

Belfkira, R., Barakat, G., & Nichita, C. (2008). Sizing Optimization of a Stand-Alone Hybrid Power Supply Unit: Wind/PV System with Battery Storage, *International Review of Electrical Engineering (IREE)*, Vol. 3, No. 5, pp. 820-828.

Billinton, R. (1984). *Power system reliability evaluation*, Plenum Press, New York.

Billinton, R., & Allan, R.N. (1992). *Reliability Evaluation of Engineering Systems: Concepts and Techniques* (2nd edition), Plenum Press, New York.

Borowy, B. S., & Salameh, Z. M. (1996). Methodology for Optimally Sizing the Combination of a Battery Bank and PV Array in a Wind/PV Hybrid System, *IEEE Transactions on Energy Conversion*, Vol. 11, No. 2, pp. 367 – 375, ISSN 0885-8969.

Chedid, R., & Rahman S. (1997). Unit Sizing and Control of Hybrid Wind–Solar Power Systems, *IEEE Transactions on Energy Conversion*, Vol. 12, No. 1, pp. 79-85, ISSN 0885-8969.

Ekren, O., Ekren, B. Y. (2008). Size Optimization of a PV/Wind Hybrid Energy Conversion System with Battery Storage Using Response Surface Methodology, *Applied Energy*, Vol. 85, No. 11, pp. 1086–1101.

Garcia, R. S., & Weisser, D. (2006). A Wind–Diesel System with Hydrogen Storage Joint Optimization of Design and Dispatch, *Renewable energy*, Vol. 31, No. 14, pp. 2296-2320.

Hakimi, S. M., & Moghaddas Tafreshi, S.M. (2009). Optimal Sizing of a Stand-Alone Hybrid Power System via Particle Swarm Optimization for Kahnouj Area in South-East of Iran, *Renewable energy*, Vol. 34, No. 7, pp. 1855-1862.

Jahanbani, A., Ardakani, F.F., & Hosseinian, S. H. (2008). A Novel Approach for Optimal Chiller Loading Using Particle Swarm Optimization, *Energy and Buildings*, Vol. 40, No. 12, pp. 2177–2187.

James A. Momoh. Smart Grid Design for Efficient and Flexible Power Networks Operation and Control,

Karki, R., & Billinton, R. (2001). Reliability/Cost Implications of PV and Wind Energy Utilization in Small Isolated Power Systems, *IEEE Transactions on Energy Conversion*, Vol. 16, No. 4, pp. 368–373, ISSN 0885-8969.

Kashefi Kaviani, A., Riahy, G.H., & Kouhsari, SH.M. (2009). Optimal Design of a Hydrogen-Based Stand-Alone Wind/PV Generating System, Considering Component Outages, *Renewable Energy*, Vol. 34, No. 11, pp. 2380–2390.

Khan, M. J., & Iqbal, M. T. (2005). Pre-feasibility Study of Stand Alone Hybrid Energy Systems for Applications in Newfoundland, *Renewable Energy*, Vol. 30, No. 6, pp. 835-854.

Koutroulis, E., Kolokotsa, D., Potirakis, A., & Kalaitzakis, K. (2006). Methodology for Optimal Sizing of Stand-Alone Photovoltaic/Wind-Generator Systems Using Genetic Algorithms, *Solar Energy*, Vol. 80, No. 9, pp. 1072–1088.

Lopez, R. D., & Agustin, J. B. (2008). Multi-Objective Design of PV-Wind-Diesel-Hydrogen-Battery Systems, *Renewable Energy*, Vol. 33, No. 12, pp. 2559-2572.

Nomura, S., Ohata, Y., Hagita, T., Tsutsui, H., Tsuji-Iio, S., & Shimada, R. (2005). Wind Farms Linked by SMES Systems, *IEEE Transactions on Applied Superconductivity*, Vol. 15, No. 2, pp. 1951–1954, ISSN 1051-8223.

Saint, B. (2009). Rural distribution system planning using smart grid technologies, pp. B3 - B3-8, ISBN 978-1-4244-3420-6

Santacana, E., Rackliffe, G., Tang, L., & Feng, X. (2010). Getting Smart, With a Clearer Vision of the Intelligent Grid, Control Emerges from Chaos, *IEEE power & energy magazine*, pp. 41-48.

Sarhaddi, F., Farahat, S., Ajam, H., Behzadmehr, A., & Mahdavi Adeli, M. (2010). An improved thermal and electrical model for a solar photovoltaic thermal (PV/T) air collector, *Applied Energy*, n. 87, pp. 2328–2339.

Shahirinia, A. H., Tafreshi, S. M., Gastaj, A. H., & Moghaddomjoo, A. R. (2005). Optimal sizing of hybrid power system using genetic algorithm, *International Conference on Future Power Systems*, pp. 6, ISBN 90-78205-02-4.

Terra, G. La, Salvina, G., & Marco, T.G. (2006). Optimal sizing procedure for hybrid solar wind power systems by Fuzzy Logic, *IEEE Conference on Mediterranean Electrotechnical* , pp. 865-868, ISBN 1-4244-0087-2.

Tina, G., Gagliano, S., & Raiti, S. (2006). Hybrid Solar/Wind Power System Probabilistic Modeling for Long-Term Performance Assessment, *Solar Energy*, Vol. 80, No. 5, pp. 578–588.

Wang, L., & Singh, Ch. (2009). Multicriteria Design of Hybrid Power Generation Systems Based on a Modified Particle Swarm Optimization Algorithm, *IEEE Transactions on Energy Conversion*, Vol. 24, No. 1, pp. 163 – 172, ISSN 0885-8969.

Wang, L., & Singh, Ch., (2007). PSO based multi-critia optimum design of a grid connected hybrid power system with multiple renewable sources of energy, *Proceedings of the 2007 IEEE Swarm Intelligence Symposium (SIS)*, pp. 250 - 257 ISBN 1-4244-0708-7.

Xu, D., Kang, L., Chang, L., & Cao, B. (2005). Optimal Sizing of Stand-Alone Hybrid Wind/PV Power Systems Using Genetic Algorithms, *In Canadian Conference on Electrical and Computer Engineering*, pp. 1722-1725, ISSN 0840-7789.

Yang, H., Lu, L., Zhou, W. (2007). A Novel Optimization Sizing Model for Hybrid Solar–Wind Power Generation System, *Solar Energy*, Vol. 81, No. 1, pp. 76–84.

Permissions

The contributors of this book come from diverse backgrounds, making this book a truly international effort. This book will bring forth new frontiers with its revolutionizing research information and detailed analysis of the nascent developments around the world.

We would like to thank Dr. Majid Nayeripour and Mostafa Kheshti, for lending their expertise to make the book truly unique. They have played a crucial role in the development of this book. Without their invaluable contribution this book wouldn't have been possible. They have made vital efforts to compile up to date information on the varied aspects of this subject to make this book a valuable addition to the collection of many professionals and students.

This book was conceptualized with the vision of imparting up-to-date information and advanced data in this field. To ensure the same, a matchless editorial board was set up. Every individual on the board went through rigorous rounds of assessment to prove their worth. After which they invested a large part of their time researching and compiling the most relevant data for our readers. Conferences and sessions were held from time to time between the editorial board and the contributing authors to present the data in the most comprehensible form. The editorial team has worked tirelessly to provide valuable and valid information to help people across the globe.

Every chapter published in this book has been scrutinized by our experts. Their significance has been extensively debated. The topics covered herein carry significant findings which will fuel the growth of the discipline. They may even be implemented as practical applications or may be referred to as a beginning point for another development. Chapters in this book were first published by InTech; hereby published with permission under the Creative Commons Attribution License or equivalent.

The editorial board has been involved in producing this book since its inception. They have spent rigorous hours researching and exploring the diverse topics which have resulted in the successful publishing of this book. They have passed on their knowledge of decades through this book. To expedite this challenging task, the publisher supported the team at every step. A small team of assistant editors was also appointed to further simplify the editing procedure and attain best results for the readers.

Our editorial team has been hand-picked from every corner of the world. Their multi-ethnicity adds dynamic inputs to the discussions which result in innovative outcomes. These outcomes are then further discussed with the researchers and contributors who give their valuable feedback and opinion regarding the same. The feedback is then collaborated with the researches and they are edited in a comprehensive manner to aid the understanding of the subject.

Apart from the editorial board, the designing team has also invested a significant amount of their time in understanding the subject and creating the most relevant covers. They scrutinized every image to scout for the most suitable representation of the subject and create an appropriate cover for the book.

The publishing team has been involved in this book since its early stages. They were actively engaged in every process, be it collecting the data, connecting with the contributors or procuring relevant information. The team has been an ardent support to the editorial, designing and production team. Their endless efforts to recruit the best for this project, has resulted in the accomplishment of this book. They are a veteran in the field of academics and their pool of knowledge is as vast as their experience in printing. Their expertise and guidance has proved useful at every step. Their uncompromising quality standards have made this book an exceptional effort. Their encouragement from time to time has been an inspiration for everyone.

The publisher and the editorial board hope that this book will prove to be a valuable piece of knowledge for researchers, students, practitioners and scholars across the globe.

List of Contributors

António C. Marques and José A. Fuinhas
NECE, University of Beira Interior, Management and Economics Department, Covilhã, Portugal

Mashauri Adam Kusekwa
Electrical Engineering Department, Dar es Salaam Institute of Technology, Dar es Salaam, Tanzania

Hasan Saygın
Istanbul Aydın University, Engineering and Architecture Faculty, Turkey

Füsun Çetin
Istanbul Technical University, Energy Institute, Turkey

Kwok W. Cheung
Alstom Grid Inc., USA

Somrat Kerdsuwan and Krongkaew Laohalidanond
The Waste Incineration Research Center, Department of Mechanical and Aerospace Engineering, King Mongkut's University of Technology North Bangkok, Thailand

Juraj Altus, Michal Pokorný and Peter Braciník
University of Žilina, Slovak Republic

Tran Tuyen Quang and Yutaro Takahashi
Department of Mechanical Engineering, Faculty of Engineering, Kyushu University, Japan

Yusuke Shiratori
Department of Mechanical Engineering, Faculty of Engineering, Kyushu University, Japan
International Institute for Carbon-Neutral Energy Research (WPI), Kyushu University, Japan

Shunsuke Taniguchi
Department of Mechanical Engineering, Faculty of Engineering, Kyushu University, Japan
International Research Center for Hydrogen Energy, Kyushu University, Japan

Kazunari Sasaki
Department of Mechanical Engineering, Faculty of Engineering, Kyushu University, Japan
International Institute for Carbon-Neutral Energy Research (WPI), Kyushu University, Japan
International Research Center for Hydrogen Energy, Kyushu University, Japan

Vorachatra Sukwattanajaroon and Suttichai Assabumrungrat
Chulalongkorn University, Faculty of Engineering, Department of Chemical Engineering, Thailand

Sumittra Charojrochkul
National Metal and Materials Technology Center, Thailand

Navadol Laosiripojana
King Mongkut's University of Technology Thonburi, Thailand

Worapon Kiatkittipong
Silpakorn University, Department of Chemical Engineering, Thailand

Chaouki Ghenai
Ocean and Mechanical Engineering Department, Florida Atlantic University, United States of America

Valentin Dogaru Ulieru1, Horia Andrei and Traian Ivanovici
Valahia University of Targoviste, Romania

Costin Cepisca
University Politehnica of Bucharest, Romania

Fatemeh Jahanbani and Gholam H. Riahy
Electrical Engineering Department, Amirkabir University of Technology, Iran

Printed in the USA
CPSIA information can be obtained
at www.ICGtesting.com
JSHW011439221024
72173JS00004B/871